THE HOUSE OF TOMORROW

THE HOUSE OF

Jean Thompson

TOMORROW

PERENNIAL LIBRARY
HARPER & ROW, PUBLISHERS
NEW YORK, EVANSTON, SAN FRANCISCO, LONDON

A hardcover edition of this book is available from Harper & Row, Publishers, Inc.

The epigraph on page ix is reprinted from *The Prophet* by Kahlil Gibran with the permission of the publishers, Alfred A. Knopf, Inc. Copyright 1923 Kahlil Gibran; renewal copyright 1951 by Administrators C.T.A. of Kahlil Gibran Estate and Mary G. Gibran.

THE HOUSE OF TOMORROW. *Copyright © 1967 by Harper & Row, Publishers, Incorporated. All rights reserved. Printed in the United States of America. No part of this book may be used or reproduced in any manner without written permission except in the case of brief quotations embodied in critical articles and reviews. For information address Harper & Row, Publishers, Inc., 10 East 53d Street, New York, N.Y. 10022. Published simultaneously in Canada by Fitzhenry & Whiteside Limited, Toronto.*

First PERENNIAL LIBRARY edition published 1974.

STANDARD BOOK NUMBER: 06-080322-3

*This book is gratefully dedicated
to the staff and volunteers
at the Booth Memorial Hospitals.*

# THE HOUSE OF TOMORROW

Your children are not your children.
They are the sons and daughters of Life's longing for itself.
They come through you but not from you,
And though they are with you yet they belong not to you.

You may give them your love but not your thoughts,
For they have their own thoughts.
You may house their bodies but not their souls,
For their souls dwell in the house of tomorrow,
 which you cannot visit, not even in your dreams.
 —From THE PROPHET, by Kahlil Gibran

PROLOGUE

For nearly six years I kept a large brown manila envelope hidden in the bottom of a chest in our storeroom. Sealed in that envelope was my diary for the better part of a year—nearly a hundred typewritten pages and eight notebooks scribbled full. Those notebooks were just small enough to fit into the pocket of my maternity smock.

I spent much of that year in a Salvation Army home for unwed mothers. We were sixty girls with a common secret, hiding from the outside world and from each other. We were called by our first names and only the first letter of our last name, and for most of us even that name was fictional.

That is one reason why I've waited this long before taking out that old diary. Our secret was a shared one, and even though I want to relate mine, I have no right to tell about theirs, so I have changed names, dates, places and physical descriptions. I've also added some background information and cut out some of the repetitions, but basically this is a true account of what I saw and felt.

Much of what I wrote then, particularly in the first part of the diary, seems sadly unrealistic and childish now. But that was the way I was, and I have tried not to change any of it.

July, 1966

SEPTEMBER

Wednesday the 2nd

A blank page in my typewriter—I can tell you, can't I? You won't give me away?

I've been holed up in this messy apartment for nearly two weeks, not even answering the phone. I'm sick and tired of my own voice talking to the walls and the books. Arguing with myself, getting nothing settled. How did I ever get into this mess? How do I get out of it? I can't even think straight anymore. I've slept more these last two weeks than ever before in my whole life. But then I wake up and there it is again, that horrible fact. I don't even want to put it on paper. I keep pretending it'll go away if I don't admit it.

I am pregnant. I am going to have a baby! So, there it is! Now talk back to me, you innocent piece of paper, just staring back at me with my own silly words.

I think I'm going crazy. I've got to talk to someone. But who? The gang is out of town still; they won't be back till school starts, and maybe that's just as well. If they know, next thing someone would tell Gene and I'd really be in trouble.

Maybe I ought to call Dorothy. I think she can keep a secret even if her husband is my teacher.

Later

God, I'm so confused. I wanted to call Dorothy, and hearing her voice was so wonderful it made me cry and she could hear I was upset. She sounded real worried and said, "Where have you been? I've called you several times and no one answered." I tried to make my voice carefree and light and I know I failed completely.

"What's wrong?" Dorothy sounded serious and very motherly all of a sudden, and I could have kicked myself for whimpering like a kid and telling her I wanted to see her.

"I'll be right over," she said, and hung up before I could say anything else. And now what do I do? Why did I have to use that stupid phone? Why did I think I could trust Dorothy? She'll write my mother, I'm sure. If she tells her husband I'll be kicked out of school. Not that it matters right now—I won't be able to attend this semester anyway.

But I've got to keep this from Mother and Dad. That's been just about the only stroke of luck in this whole mess, the fact that Daddy is spending this year as an exchange professor in Europe. And my kid brother is in the Air Force in Germany. I'll have to tell them before they come back in June, of course, but by then I'll be straightened out. The baby will be here and we'll be on our own someplace.

Right now I've got to keep it from them. I can just see Mother flying home from France all upset. I can't stand thinking of all those tears and all that "shame, shame" business. And Daddy— it would just kill him. Not that he's approved of anything else I've done for the last several years.

This whole thing is a perfect nightmare. Sometimes I manage to think it's all a giant hoax. But of course I know it's true. It's crazy but I've known it ever since that last night with Gene. Deep down I've known it all along. Of course, I played the role. I counted the days in spite of myself, and went to a doctor just so I could tell Gene I had and give him that lie about everything being O.K.

How confused can a person get? I went for long vigorous walks and swam straight out into the ocean to the point of exhaustion, and like a coward thought maybe a big wave would come or a strong current, taking it all out of my hands. . . . I even prayed for a miracle.

"Dear God, please let me start menstruating and I promise never to take another chance." Deep inside, something in me knew that prayer wasn't meant to be answered.

What if I were dead? I've thought about it and I'll probably think about it again. So simple, no more problems. Nobody embarrassed and ashamed. Just keep on swimming. That's how I would do it, I've thought. But I don't have the guts to do it. I keep thinking that even if I might have the right to choose death for myself I don't have the right to kill the baby. That would be murder. Just like abortion would be murder. That's already another human being there inside me, even if he or she is only about the size of a walnut.

Thursday the 3rd

I'm installed in Dorothy's house. She said I needed to get away from that dreary apartment. I can stay here for maybe two weeks till it's time to enroll for the fall semester. By then I'll have to have some concrete plans.

Of course, I told Dorothy everything yesterday when she came to my apartment. I think I had expected her to act shocked or hurt or something like that, but instead she listened very calmly and sipped my lousy coffee. I could see from the way her blue eyes looked all over me and my room that she was disgusted with the overflowing ash trays and the stacks of books and clothes everywhere and the dust so thick every surface looked gray. She'd never been there before and I should have cleaned it up before she came. It really didn't look too bad when I had candles burning in the Chianti bottles on the orange crate I used for a table, with the light flickering over the olive-green walls and the Toulouse-Lautrec print of the absinthe drinkers. But yesterday afternoon, with the light filtering in through those dark burlap curtains, the place looked pretty bad. I could see how someone could be alarmed.

It's funny, but I had never noticed just how much Dorothy looks like my mother. Yesterday it really hit me. They both have large blue eyes, a strong face with broad cheekbones, and light-brown hair graying and piled in a soft bun. They are just about the same age. But inside they are as different as day and night. Dorothy is always an easy listener, her house open to anyone who wants to drop by for a piece of homemade pie and a chat. You know she really cares because she'll let you know immediately if she doesn't agree with what you think and do. Not condemning or judging, just letting you know her opinion, take it or leave it.

My mother is a professor's wife too, but I don't think I've had a real talk with her for years. She is sweet and all that, but she's as narrow-minded as they come. You know—made up her mind years ago and nothing can change it. She talks a lot about helping people in need—like her church work and her charities. But I don't think she ever stopped to think that what people really want is just plain understanding.

Dorothy asked me what I was going to do with myself.

"Go away, I guess," I said, "to a big city in the Midwest, maybe. Stay there till the baby gets here and we can live there together." I could hear how silly it sounded, but I got mad when

Dorothy smiled and said very calmly, "All alone?" I felt she was deliberately forcing me into a corner.

"And why not?" I said. "I've had child-guidance courses and worked as a camp counselor." My voice was rising. "Lots of girls get married and have babies while they're still in high school—"

"They get married," Dorothy interrupted me. I got her point. But of course it's too late to think about that now. My situation isn't perfect—she doesn't need to remind me of that—but I think I'm more qualified to raise my child than some mothers I've seen. It wasn't much use trying to explain that to Dorothy. Besides, I had an awful headache and for some reason I couldn't keep from crying.

Dorothy let me cry and then she looked straight at me and said, "Do what is right for the two of you—but get out of this hole you've dug for yourself." She meant my messy apartment. "Get out and start functioning. Meet people, get a job, get ready for the baby. . . ."

I let her talk. It sounded so easy to hear her say those things, and one day I'll have to think about them. But not just yet. I'm just too tired now.

Dorothy's youngest son is in the Navy. I'm staying in his room.

Friday the 4th

I've been reading a book on the psychology of the unwed mother. I saw it downstairs and couldn't help picking it up. I guess the house of a sociology professor isn't the best place to stay, if you are expecting a baby out of wedlock and would rather not think about it for a while.

The book says such a pregnancy is rarely accidental. It says the girl nearly always wants it—as a crutch, an excuse to fail, a way to rebel or demonstrate against her parents or other authority she felt she never could free herself from. Also the book says unmarried mothers often keep their babies because they are emotionally immature and unable to face reality. Phew, that sounds like a mouthful, as if the author is really looking for symptoms where there aren't any. That couldn't possibly be all true. Of course, maybe in some instances, but not in mine.

I admit I was beginning to make bad grades. But I couldn't have been afraid of failure; my teachers always said I should apply myself, that I am bright enough. I made straight A's through my sophomore year but then school started making less and less sense to me. In a crazy way I'm almost glad I don't have to go to school next year. I need time to think things over, decide what I want to do with myself. I thought I might want to go on and get a master's degree in psychology and be a social worker. Do something about this mixed-up world we live in. The way things are—one little war after another, riots, and millions starving in India and all that—I don't know what there is to believe in.

I guess Daddy thinks I'm a lost cause. He thinks I don't know who I am and where I'm going. Hell, he's the one who wanted me to come here instead of going to college at home. For once we agreed. I don't think I could stand going to school where he is teaching. What used to bug me was the way his students used to tell me how lucky I was to have a father like him. They said he gave such good advice and always had time to listen. Little did they know that whenever we got into a discussion about anything at home it always turned into an argument. Daddy thought I ought to be number one in my class, and he thought I brought absolute disgrace on the family because my friends grew beards and wore sandals and played guitars.

I think my friends are real people. They are natural because they don't want to pretend. Why should we wear make-up to look like something we're not? Why should we cut our hair when obviously it keeps growing? So Daddy called it childish and immature—a failure to cope with realities. Because we don't make believe and pretend all is well with a phony world.

By that standard, maybe I am a failure the way the book says. But I didn't want to get pregnant. If I did, I'd be happy as a lark now, wouldn't I, instead of in a mess?

Sunday the 6th

I've written Mother and Dad about not going back to school. They probably knew it was coming. I said I needed time to think about my future and I didn't want to waste their money on tuition when I'm not serious about studying. That part is true. I've hated taking their money. They've still got my kid brother Tommy to put through school when he gets out of the Air Force two years from now. He'll be well worth their efforts, I'm sure. He's always been the conscientious one in the family.

Next week is registration for the fall semester. I've got to do something. Dorothy came in and turned off my favorite TV program this morning just to ask me what I've done about the baby. Of course she knows darn well I haven't done a thing. She called me a selfish, self-centered brat and for the first time I saw her really angry.

"You sit around brooding about yourself!" she said. "Can't you see how cruel and selfish it is to bring that baby into the world without getting a name for it?"

I've never seen her carry on like that. I didn't know what to say, but I knew she would calm down and be sorry about the whole thing later. I know she doesn't really think those things about me. She knows I've been nauseated, and after all I'm in a pretty horrible situation. She can't really expect me to settle everything this soon. I told her I was sick to my stomach, and for a second I thought she was going to slap me.

"So what!" she said. "I've had four kids and been a lot sicker than you. Quit thinking about yourself and go get a job!"

Late evening

Dorothy came to my room and said she was sorry.

"You know I want to help you," she said, "but you've got to help yourself first. The baby is your responsibility." She sat on my bed and waited for me to say something, but I didn't.

"Why don't you want to marry Gene?" I had been waiting for that question.

"You know he's married," I said, and I was a little angry with her for even asking.

"You told me he was separated from his wife when you met him," she said. "You said he wanted to marry you. I'm sure he would if he knew you were going to have his baby. Why don't you tell him?"

"I just don't want to mess up his life," I said, and she shrugged her shoulders.

"O.K. That's your business," she said. "But if you're going to keep that baby you should marry someone else—at least on paper. Gene will probably put two and two together if he ever hears about you living alone with a child." That caught me off guard and my voice wasn't too steady when I told her I was going far away and I'd never see Gene again.

"There is a right thing for you to do," Dorothy said softly, "and I know you can find it. But you must start looking."

Yes, I've got to start looking. But where? To Gene? No, I've already burned that bridge and it wouldn't be fair to go back. I think he loves his wife. I know he loves his kids. He'd never be happy with me, or I with him, for that matter. That trite thing about living in two different worlds really holds for us.

He looked so miserable and lost the night he came into the café near Harvard Square where our gang usually hangs out. Gene looked completely misplaced there in his business suit, but he sat down at a table by himself and after a while the kids stopped staring at him.

Joe Farrari was reading some of his own poetry and I was making background music on the guitar. I felt the stranger looking at me. All the way through our program those eyes were there and I really felt them all through me. As if we were really close, as if we understood each other. When we were through I just went down and sat next to him. I knew he wanted to talk to me. It's funny, but when I look back on it I see that we were never again as close as we were those first couple of hours when we were talking with our eyes.

Gene walked me home that night, leaving his Cadillac parked near the square and carrying my guitar over his shoulder. He held my hand and we walked along the river talking. That is, he did most of the talking. I can be a good listener when I want to.

Now I think it didn't matter too much *who* listened—he just needed to tell someone. Someone different. So he talked about working his way through high school and going into the Navy, and then starting in business for himself and getting married when he was only eighteen, his wife working all the time. Now they own a chain of laundromats and a ski lodge in New Hampshire, and have a maid and go to Europe on their vacations.

"I've got all the things I worked so hard for," he said, squeezing my hand. "And now I'm thirty-five years old, and suddenly it isn't important anymore. Nothing is. I've got to find myself again."

In my apartment that night we made love by candlelight and listened to Joan Baez. Gene had never even heard of her!

That was in April, and the ice was breaking on the river. All spring we did things together. Gene was like a big kid on his first vacation. He'd never fed the ducks in the pond or gone on picnics in the woods or sat on the grass listening to symphony under the stars. He hadn't browsed through secondhand bookstores or little art galleries or sailed paper boats on the river. I thought I was really falling in love this time.

Till one day in June he proposed to me, and I didn't get happy. I got scared. He said he loved me because I was so honest and didn't pretend to be anything but myself. But I wasn't honest enough to tell him I didn't want to live in a nice house with two cars in the garage and four kids. I wasn't even honest enough to admit it to myself at first, but I was becoming more miserable all the time. I talked about his kids a lot because I knew how much he loved them, and getting a divorce to marry me would mean giving up those kids. I could tell he thought about that, too.

When I thought maybe I was pregnant I got frantic at the idea of being trapped with him for the rest of my life, so one day I told him he ought to go back to his wife and make a try of it—for the kids' sake, I said.

"How can you be this unselfish?" he said, and I felt like a heel and only smiled. When he kissed me good-bye I felt like someone who's just been saved from the rip tide.

So there it is. I've written it down now. I'm scared to death of marriage, marriage to anyone. I'm scared of being trapped. But I am going to have a baby and I'm going to keep it. I'm not scared of motherhood.

But what about a name for my baby? Dorothy is right about that. Married on paper—to whom? I can't just stop someone on the street and say, look, I want to raise a baby without a father, how about lending me your name?

Maybe if I tell someone I can trust, someone who can keep a secret, maybe he'd offer to help me? David might be the type, and I've known him all my life. Maybe I'll write him tomorrow.

Monday the 7th

I told Dorothy I've written to David in New York. She said fine.

"But remember at first you wanted to raise the baby without any legalities. Maybe you're still out to demonstrate."

"I've got nothing to demonstrate about," I said.

"Then maybe you want an excuse to drop out of school."

Dorothy makes me mad sometimes. She has the nerve to ask the strangest questions. She knows I can't hold down a job, raise a child and complete an education all at once. Why does she remind me of the things I've got to give up? My first duty is to that child. Any mother would agree.

Wednesday the 9th

David must have answered right away. The letter got here this morning, airmail. It scared me a little. I didn't want to open it. As long as it was sealed I could hold it in my hand and imagine what was in it.

I understand how you wanted to keep the child at first. I am confident you already see how impossible that solution is. You don't have what the baby needs most, a home and a husband.

Also, you are an intelligent girl with a fine career ahead of you.

This experience, I am sure, will give you maturity and greater understanding. You will have much to offer when all this is over. . . .

This may not be the kind of letter you're looking for but, frankly, I'm a little stunned by the whole thing. You should have known better.

I'm glad you found me worthy of your confidence. Let me know what you will do; I am curious. Will you go to a home for unwed mothers? Use agency or independent adoption? After all, this is one experience I can't have at first hand.

Let me hear from you as time goes on. Good luck. I'm sure you can make the best of this.

Love,
David

I felt numb. There was a spot of grease on the stove and I cleaned it. The water was boiling and I made some tea. I counted the five cigarettes in my package and lit one of them. I kept my mind blank and only thought that I ought to iron a skirt to wear tomorrow. And then the tears came. Bits of thoughts whirled in my mind, and crying made sense. Dorothy walked in and I handed her the letter. She read it and said, "Well?"

"I feel so cheap," I cried. "I didn't mean it that way."

"I don't think he thought so either," said Dorothy. She had brought me a tiny white pill.

"This'll make you sleep awhile," she said. Thank God I didn't dream.

Thursday the 10th

This morning I almost fainted in the shower and Dorothy sent me back to bed.

"You'd better be careful. We don't want you to have a miscarriage, do we?" For some reason that sounded so funny to both of us that we laughed till our stomachs ached and we had tears in our eyes.

Friday the 11th

Registration starts at school Monday, and Joe Farrari called me today. He just got back in town and tried to reach me at the apartment. The landlady told him where I was.

"How about making the town tonight?" he asked, and I had a wild thought. Maybe Joe can help. He's always been a regular guy.

"O.K.," I said. "Let's go to the beach."

I don't want to think about David's letter or marriage or having a baby. Maybe I can just have fun tonight—the way I did in the old days. And maybe I can tell Joe what's going and maybe he'll say something. . . .

Late Friday

It was cool on the beach and dark clouds hid the stars. The sound of the surf drowned our voices and the crests of the waves gleamed frosty blue. I love the sea but breaking waves are scary sometimes; they keep coming and you can't stop them.

"What's the matter?" Joe said. He slid his arm around my shoulders. "You were never this quiet. Got troubles?" I felt my throat tighten and I was getting hot, the way I do when I know I'm going to say something I shouldn't.

"I'm pregnant." I watched his face fall into a stupid stare.

"I'll be damned," Joe said and slid back behind the wheel. "Who's the guy—are you getting married?"

"No," I said. "And you never met him."

"I'll be damned," he said again and slid back toward me. The surf was breaking in my ears.

"I'm sorry, kid," Joe said, and then he kissed me. He'd never kissed me before and I closed my eyes, holding onto him, wanting to be near someone and not alone. And then his hands tore at my clothes and he was breathing heavily. Something had gone wrong. I tried pushing him away but he didn't even notice. Then I bit him hard and he sat up with a jerk.

"What the hell!"

"Please," I said. "Don't be like that."

"Geeze," he said. "You don't have to put on a show for an old friend."

"We're not that kind of friends." I wanted to cry.

"You're pregnant, aren't you?" He was angry. "So what's the worry, let's have some fun." I wanted to die, but he was right. I was getting what I deserved.

"Take me home!" I was sobbing, and I guess Joe hadn't expected me to cry. He didn't say a word all the way home.

Saturday the 12th

I'm leaving. There is no use putting it off anymore. I told Dorothy this morning and she said, "That stinker," meaning Joe. But his reaction was a normal one, and I can expect others to react the same way. I can't run around throwing my problem into other people's laps and expect them to solve it for me. I've got to face it alone.

"I'll go to California," I said. I've read that California is one of the states where they don't mention the mother's marital status on the birth certificate. Dorothy looked at me, and I remembered that afternoon in my apartment. It seemed ages ago and it's only a couple of weeks. I wanted to cry on her shoulder again.

"What will you do there?" she said.

"Find a home for unwed mothers," I said. "There's bound to be one or more in the big cities." I wanted to smile in spite of it all. "You know, California, land of opportunity, where you can win fame or get lost in a day."

"Why don't you call the Salvation Army?" Dorothy said. "They would have that kind of information." She looked at me in her most matter-of-fact way. "Let me know if you need help in finding adoptive parents for the baby."

"No," I said. "I don't know what I'll do, but I'm not giving up the baby."

Sunday the 13th

Dorothy helped me pack the few things I am taking along. We've stored most of my clothes and books in her attic. She'll receive my mail here and forward it. I've told Mother and Dad that I'll be staying here this year, that Dorothy has invited me since her children have all gone now. I think we have worked out the practical details pretty well. If any of my friends calls me here, Dorothy will tell them I'm out of town for a couple of days.

It's almost time to go. Dorothy will drive me to the station. She's been so wonderful, but I guess this is the point where I go on alone.

Sunday night

The train isn't very crowded. I've got a double seat to myself, by the window. Can't afford a Pullman, but then I don't think I'll sleep very much anyway. Outside the landscape flies by—dark clusters of trees and hills outlined against the evening sky—

houses with windows showing families around the supper table under the lamplight.

I am alone, but this kind of loneliness doesn't matter. I feel snug and safe inside the train. It moves me from one end of the country to another. The tracks stretch ahead, the engineer knows what he's doing and I don't have to worry about anything.

Across the aisle and a few seats ahead of me is a sailor asleep, with his white cap over his eyes. Maybe he's going home on leave.

I once wanted to be a boy. When I was little I played with the boys, and I could run as fast as they did and climb the tallest tree. I didn't like dolls and girls, and my father used to laugh and call me his little tomboy. I think maybe he wanted me to be a boy. I used to think boys could do things girls couldn't do. Maybe if I'd been a boy I might have been a sailor now, asleep in a train on my way to someplace. Instead, I've got a real little live human being growing inside me. In spite of everything, I'm glad I'm a girl!

Later

The sailor's name is Dean Kirkpatrick but his friends call him Patrick because of his red hair, he said. I met him in the diner— I was sitting there with my pot of coffee, staring at the big yellow moon, and almost felt sorry for myself. He was suddenly standing next to my table with a big friendly grin on his face, his white cap in his hand, asking if I would mind if he sat down. I thought, why not, he's a stranger, he'll never know where I'm going.

"Please do," I said, and turned to stare out the window again. Our reflection looked as if we were together. He was looking at the window too. He smiled and nodded, and said, "Nice-looking couple, aren't they?" And I couldn't help laughing.

"It's fun," he said, looking directly at me across the table. "A trainful of people going someplace from somewhere." He laughed and looked as if he was having a ball. "Some of them think they'll be happier where they're going, some think they left happiness behind—and some don't think at all. They just go." His face looked suddenly serious but his eyes were still full of fun.

"I'm not really nuts," he said. "I just happen to be a compulsive talker—especially on trains—and especially to pretty girls." The way he said it I couldn't feel embarrassed or anything, I couldn't keep from laughing with him.

"Meeting people on trains is like peeking in the middle of books," he said. "You don't know the beginning and you won't know the end. You know just enough to be interested and not enough to be disappointed." For a quick moment I wondered what he would say if I told him the end of my "book," but he kept on talking, chasing away that awful feeling of being an outsider and watching people talking and laughing and living normal nonsecret lives. He is on his way home after three years in the Navy and he's got money saved to finish college.

"Isn't it great to be alive?" he said, and I nodded at him even though I felt something like envy. Yes, it must be great to be alive when you're on top of the world the way Patrick feels now, when you haven't got yourself into a great big mess. But I'm glad he's on the train. That smile of his is contagious. And he likes me, he really does. I can see it in his eyes and it feels good to have someone tell you silly, funny compliments. He's going to the Coast too. That'll make the three days go quicker.

Wednesday the 16th

We'll be there tomorrow—then what? But I still have a few more hours when I don't have to think about it. The trip has gone so fast, mostly because of Patrick. He is going to get a degree in psychology and be a counselor, he says. He'll be a good one, I'm sure. Today he "analyzed" me. He said I'm bright, warm and adventurous ("Or you wouldn't have listened to a stranger like me"), but also that I'm lonely and don't have enough confidence in myself.

"You've got a wonderful life waiting for you," he said. "But you've got to go out there and live it, not hide in a corner."

"Thank you doctor," I said, and we both laughed. I think that if I told him where I'm going he would understand, maybe even want to see me again sometime, but I don't dare. I've felt almost happy these last couple of days. I don't want to talk about what's ahead; it'll come soon enough.

Maybe it won't be so bad. This secrecy will only have to last a few months and then there will be a whole wonderful life ahead. And a baby to share it with.

Thursday the 17th

I said good-bye to Patrick at the station and explained why I can't see him again while I'm here. I said I would be visiting relatives for just a few days before going back to Boston. I gave him Dorothy's address, and hope he'll write me even if he's sore about not seeing me here.

Then I took a taxi to the Y.W.C.A. and registered as Jean Thompson from Chicago. I've never been in Chicago except for a stopover at the station or the airport. If I'm going to lie about it I might as well pick someplace I don't know a thing about. I like my new name. Jean Thompson sounds like a real person.

SEPTEMBER

Saturday the 19th

This is the end of the line, no more stalling. I am broke. Now is the time to take the next step.

I woke up early this morning because the girl in the bed next to mine was crying. She looked a mess, blond hair tangled, eyes swollen red—poor thing. She's been in this country from England for only a month and she's already been involved with a smart guy from Hollywood, who took her for a two weeks' "honeymoon" without a wedding and left her here while he took off for Vegas with her last hundred dollars. She told me the whole story in between sobs, how wonderful that guy was, how she was afraid maybe he wouldn't get his divorce after all, and maybe she's pregnant. I promised I wouldn't tell a soul about that, of course.

"Now I can't afford to stay here any longer, and how is he going to find me? He promised to call." She looked at me as though I knew all the answers, and I said she could go get a job, and if she's pregnant, there is always the Salvation Army.

Then I walked down the hall to the pay phone and dialed the Salvation Army. A pleasant voice answered, "Can I help you?" And I've never meant it more than when I said, "You sure can! I'm going to have a baby and I'm not married and I'm nearly broke. I don't know what to do."

"What's your name?" said the voice at the other end, as if what I'd just said was the sort of thing the voice handled every day. I told her my new name and she said to come right over.

"Since you're broke you'll probably want to make an arrangement right away, won't you?" she said, and I realized suddenly just how frightened and lost I had felt. It was going to be all right now. I would have food and shelter and no questions asked.

I went back to the room and gave the English girl my last ten-dollar bill.

The social worker was young and pretty. A small overnight bag stood by her chair and I remembered it was Saturday. A long time ago I had gone away on week-end trips too. It seems like an eternity ago.

"I have called our Home for Unwed Mothers about you," she said. "We don't usually admit girls on Saturday, but there happens to be room. You may go right away if you want to." She smiled and tapped her pencil on the pad.

"There are some questions," she said. "I suppose Jean Thompson isn't your real name?" I must have looked surprised and she said, "That's quite common. Many girls don't give their real names. We'd like to know, of course, in case of an emergency. But it would be confidential."

"I'd rather not," I said.

"All right, then what shall I put down as your hometown?"

"Chicago." She asked my age and occupation.

"Twenty-one, college student."

"What field?"

"Sociology," I said, and she laughed.

"Doing it the hard way, aren't you?" It wasn't hard to laugh with her. She wasn't at all what I had been afraid of—someone like my mother, maybe, strict and disapproving, there to "save" me from the gutter.

We shook hands and she wished me luck. I walked to the bus depot and people with Saturday looks on their faces hurried around me. I wanted to laugh out loud. It seemed too fantastic to be true. I was on my way to a home for unwed mothers!

Saturday afternoon

The house stands on a hilltop surrounded by park-like gardens with large old trees and brilliant flower beds along winding paths. A sign at the entrance to the park says: BOOTH MEMORIAL HOSPITAL. It looks more like a school dormitory to me. There are bright-colored curtains behind the wide windows and the big two-story house looks friendly.

I climbed the steep flight of steps from the street and tried to open the big green front door; it was locked. The tight feeling in my stomach was just plain fear. I felt as if I was being watched, and when I looked up I saw the curtain move behind a window upstairs. I wondered if it was one of the girls, and what they would be like. I felt like running back down the stairs. Going through that green door would be so final. I thought: that's where it will happen, in that house—all the things I'm scared to think about, the baby, how I'll feel when it's over, what I'll do. Even as I rang the bell I wished somehow I could get out of it, go back and undo that mistake.

If the door hadn't been opened right away I might have run. But a woman with a tight, pinched face opened it as if she'd been waiting for me. Without a word, she held it open so I could walk in and then locked it behind me again, and I couldn't help feeling trapped.

"Miss Peterson at the welfare office called about you," she said, without answering my "How do you do."

"You know we don't usually admit girls on Saturday. You'll fill out your papers Monday." She marched into the office adjoining the hall and flicked the switch on the intercom. I heard a wave of noise from a room somewhere, girls' chatter and laughter; they didn't sound like desperate unwed mothers.

The frozen-faced lady called into the microphone: "Karen T. and Lucy S." And several voices answered, "Not here!" She flipped another switch and tried again: "Karen T. and Lucy S." And the voices came over the din in the room, "Coming!"

"They'll be here to get you in a moment," the woman said and turned her back on me to shuffle papers on a desk.

I sat on the olive-green plastic couch and tried to think of something pleasant. The foyer looked depressing and smelled of strong soap. Brown, slightly worn linoleum covered the floor, and the walls were green. The furniture was heavy and dark. Someone had hung bright new flower-print curtains in front of the tall narrow window and had put artificial roses in a vase on the table. I leafed through the magazines, old copies of *Life* and *The New Yorker,* and *War Cry,* of course, the Salvation Army magazine. The name always seemed silly to me.

The frozen-faced lady wasn't wearing a uniform. She couldn't be a Soldier of the Cross (not with that look on her face anyway, I thought). There was a cupboard with glass doors in the corner. On the shelves were embroidered pillowcases and baby clothes neatly folded and priced. Maybe the girls made them. I can't think of anything I'd rather do less. Two girls walked by in the hall and peeked at me through the door. They wore bright cotton smocks over their big stomachs and were laughing. How can they be so cheerful in this place? They are safe in here, of course.

New footsteps sounded through the hall and two girls came into the foyer, one a tall, chocolate-brown Negro, the other white with dark hair and sparkling blue eyes. They smiled hello and the frozen-faced lady came to introduce us.

"A new girl, Jean T. She'll be in your dorm. This is Karen T. and Lucy S. They'll show you around." I smiled and wanted to giggle. How funny, really, the way we're going to share something as real as having babies and we call each other by a phony first name and last initial only. Or maybe I ought to cry instead.

"Hi, Jean," Karen said, blue eyes smiling. "Hope you'll like it here." She caught my stare and laughed. "Really, it isn't bad. Wait and see, the girls are lots of fun." Karen and Lucy each took one of my bags. Down the hall, out of hearing range from the office, Lucy nodded over her shoulder and whispered, "Don't mind Mrs. Warren. She always scares the new ones, but she isn't bad once you get used to the fact that she never smiles. At least she's consistent."

The long hall was gloomy, brown floor, green walls, no windows. But now I could hear the sounds of the house all around us. They were busy, happy sounds—rattling of dishes somewhere, voices talking and laughing, someone whistling. A door banged, and I heard a piano and someone singing.

"We'll take you on the whole tour of the premises," Karen said with a smile.

"And don't worry if you don't remember all the names." Lucy chuckled and put a chocolate-brown finger over her full mouth. "In here we don't mind, if you know what I mean."

"The girls sleep in four dorms," Karen said. "Two downstairs and two upstairs. We're upstairs and if you don't mind we'll take your bags there first."

"Gee I'm sorry." I suddenly realized that Karen was breathing heavily and straining to carry my bag. "Let me carry it." I grabbed the bag and she let go at once, smiling gratefully.

"When you get a little more pregnant I'll show you all the respect a mother-to-be deserves," she said. "Now you don't even look as if you belong in here. When is your due-date?" I'd never heard that expression before and they both laughed at me.

"You might as well get prepared," said Lucy. "We ask two questions in here. When is your due-date and where are you from."

"That means we want to know when your baby is due and we hope you don't come from someplace we know," Karen explained.

"I'm due in March," I said, and they exchanged glances.

"It'll go fast," said Lucy quickly. "Christmas will be here before you know it."

"I guess so," I said, and the time stretched out ahead of me in months and weeks and days and minutes to be counted.

"And I'm from Chicago," I said, and Karen said, "Don't worry, we haven't got anyone claiming Chicago as a hometown right now. Lucy comes from near here and I'm from Missouri."

The steps leading upstairs are a dark green. Lucy and Karen had to stop to catch their breath several times. They looked at my flat stomach and I felt almost guilty for not being tired.

There are fifteen girls in each of the two dorms upstairs, with just one bathroom between them. I bet that gets crowded. Six lavatories, five toilets, two showers and a tub. The dorms are identical except for the color. Ours is green, the other one pink. Thank heavens—I can't stand pink. The beds feel comfortable enough. They stand, headboards against the walls, around the large square room. Fifteen narrow lockers and four large chests of drawers are lined up back to back down the center of the room. Everything is green except for the girls and the bright yellow and red curtains. I've slept in dorms like this in summer camp. You don't get much privacy, but then maybe I don't need that here. Maybe the noise and movement keep you from brooding too much.

The room was full of girls when we came in the door. They were resting on the beds, or sewing or writing letters. Two radios were playing different songs at the same time and the noise was chaotic. Lucy clapped her hands and yelled, "Quiet, girls! Meet a new one." Dead silence fell over the room at once and twelve pairs of eyes traveled up and down over my flat stomach.

"Sure you got the right address?" Several girls laughed and the redhead in the corner bed who had asked grinned proudly.

"Be good girls now and introduce yourselves," Lucy said. "This is Jean T. from Chicago."

"Hi, Jean. Welcome aboard. Nice to know you." The names and the greetings came from all sides at once and I found myself just standing there thinking that, whatever it was I had expected the girls to be like this certainly wasn't it. I mean they were just like any other girls I know, nice, friendly, open, like roommates in school.

The only difference was that everyone was obviously pregnant. Some of them had propped pillows under their big bellies to rest more comfortably, and when they got up they moved slowly and waddled when they walked. It turns out all of them except one are due long before March. The one is called Cecilie. She is due in April and looks as skinny and out of place as I do.

"Let's dump your stuff and take the tour." Karen pointed to the bed in the corner next to hers. "You'll sleep here. You've got one locker and one drawer. Monday you can put your empty bags in the storeroom downstairs." I felt very tired suddenly.

"Let's rest for a while and take the tour later," I said, and Karen flopped on her bed with a loud sigh.

"Best idea I've heard in a long time," she said. "Nobody needs to tell me twice to rest. I was born lazy."

I closed my eyes and the noises in the room didn't bother me. I thought: I'm here and that's what I've been wanting for a long time. The outside world can't touch me now. My pregnancy is the one thing I don't have to lie about in here. I've even cut my hair! I did it myself at the "Y" yesterday. I bought a lipstick too. No more longhair stuff and dungarees and sandals, not while I'm here anyway. My guitar is safely stored in Dorothy's attic. And Jean Thompson from Chicago is a sensible, average kind of girl.

Later Saturday

I slept till a bell rang. All the girls were jumping off their beds heading for the door.

"Chow time," said Karen. "The highlight of the day." We marched downstairs and through the long hall, more and more girls joining the line. The sound of our steps and our loud voices echoed in the bare hall. Then down another flight of stairs to the dining room in the basement. The line stopped in front of the closed door and Karen explained the system: first bell means we've got to hurry into line outside the dining room and then we wait till the second bell rings and march quietly to our places.

We waited in that drab hall where the heating pipes go along the ceiling. Karen told me the kitchen is down there too and the laundry room, the storage room, the beauty shop and a tiny little room where the girls are allowed to smoke. Then the bell rang and we filed in and I was glad to see that the dining room is painted white and the oilcloth on the small tables for six and eight is checkered a bright red and white. Through the windows high on the wall I could see green treetops and blue sky.

Karen steered me to a small table for four in a corner, far from the head table where a plump woman in a cook's uniform was already standing behind her chair, hands folded on the backrest. I saw Lucy heading our way with a small Oriental girl just behind her. I was glad to see a face I knew and Lucy, showing her bright teeth in the dark brown face, introduced us, "Mikki, due any time—Jean, due in March."

"Let's bow our heads, girls." The cook's voice was soft. The girls behind their chairs bent their heads, their hands folded under large and small tummies. I couldn't bend my head. I looked through the window at two birds resting on a telephone wire and hoped that God, if He was there, would understand.

"Lord, we ask Thy blessing," came that soft voice, and I felt like a hypocrite. "We are poor lost sinners and our only hope for life is through Thy Son, Jesus. In His name we ask Thy blessing on our meal. Amen."

"Amen," came a mumbling from the girls, and Karen grimaced as she sat down.

"Don't forget your pills, girls," she said. "Healthy mommies, healthy babies, you know." She flipped three pills from the little paper cup into her mouth and followed with a sip of water. Iron, vitamins, calcium, they were all there. I swallowed mine in one gulp and smiled at Karen. I think I'm going to like her, whoever she is. The food was excellent. Broiled liver (no onions) and baked potatoes. Rice pudding for dessert, my favorite. Lucy and Mikki, the Japanese girl, kept up a constant barrage of friendly kidding back and forth.

"Gonna put those crumbs in my bed?" Lucy asked when Mikki crumbled her cracker deliberately between slender fingers.

"Only because you short-sheeted me last night," said Mikki, sweetly.

"You two are going to chase each other into the delivery room one of these days," Karen said. Lucy and Mikki smiled. "That's where we're all heading," said Mikki. "We're just trying to speed the action a little."

After supper Mikki went back upstairs and I came along with Karen and Lucy for a smoke in what Lucy called the "sanctuary."

"It's the only place in the house we can smoke," she explained. "The staff can't stand to smell the stuff, so they don't come around. We can sort of be a little more free down here." The little room was crowded with girls and the stink of tobacco was almost too much, even for a regular smoker like me. There is only one small window high in the wall, looking straight out at the back steps. The window is so filthy that very little light can come through. The concrete walls look as if they've never been painted, but by now they are stained a dark gray by dirt and smoke.

A group of girls were playing poker around a rickety table under the window. Others were chatting or reading or even sewing, perched on old furniture of every style and color. Most of the chairs looked as if they needed paint and repair badly. The smoke was getting in my eyes and I had to cough. "Not much of a view you've got," I said, and Karen looked at me.

"Are you kidding?" she said. "It's a great view—we can see the doctor coming and going and we know when a girl delivers. We have sort of a worm's-eye view of the whole thing." I could see her point. Just then a car drove up to the back steps and one of the girls jumped up on her chair to see better.

"Who's in labor? It's Doctor Carlsen," she said, and one of the poker players answered, "It's Carol. She's been upstairs for several hours now." Everybody suddenly looked very cheerful and somebody yelled, "Hurrah, she made it!"

Karen explained it to me. "Carol's folks are coming to see her next week—they don't know she's having a baby."

"It often happens that way," said Lucy. "A girl gets in a really tight squeeze—but no one's been caught yet. I guess someone is watching over us." Several girls nodded. They suddenly looked quite serious. I had the odd feeling that those poker-playing girls, with their bellies full of what most churches would call babies conceived in sin, have faith!

Karen pulled my arm. "Our tour," she said. "We've got to hurry if you want to see the place before bedtime. Lights are out at ten." We looked in the door to the huge kitchen. There's a bakery, too, and a separate dining area for the housekeeping staff. Karen explained that the dining room for the Salvation Army officers and the social workers is on the first floor; directly above is the diet kitchen for the mothers' ward where the nurses and the girls on special diets eat. The food is carried upstairs from the kitchen on a special elevator.

The housekeeping staff and the social workers work here only during the day, Karen said. Two nurses are on duty on the hospital floor around the clock, and the Salvation Army officers who are in over-all charge live in staff quarters nearby. "They're very nice," Karen said. "They make you feel they really care and they don't act as if they think we're big sinners just because we're in trouble."

"What about the social workers?" I said.

"You'll be assigned to one. You can talk over your plans for the future, and she handles the adoption through the state agency."

"I'm keeping my baby," I said. Karen looked at me, then quickly turned away.

"Not many girls in here do," she said, and I was glad she let me go first up the stairs so she wouldn't see me blushing.

"On the main floor are the offices for the staff and social workers," she resumed. "We have two case workers and one group worker. She's in charge of arts and crafts, and you can pick your own hobby; paint, ceramics, sewing, leathercraft—you name it, we've got it."

"Do I *have* to do something?" I thought of the embroidered baby clothes for sale in that glass cabinet in the entrance hall.

"Of course not," Karen said. "You'll be assigned to a housekeeping job, you know, laundry, kitchen, cleaning. That's the only thing you *have* to do—except go to chapel every Sunday and Wednesday. No one gets out of that."

The chapel is just off the main hall; the door was open. The lights were off in the room and the light from a lamp outside shone through the stained-glass window. We could see the rows of pews and the large cross behind the small altar.

"It isn't bad," Karen whispered. "One of the girls can really play the organ and Cora in our dorm is the soloist. She's a tough kid—but she's got a beautiful voice."

"I thought maybe they'd try to 'save' us," I whispered. "I don't think I could take that."

Karen shook her head. "The Salvation Army officers won't do that," she said, "but watch out for the cooks and the housekeeper. They're plain soldiers in the Salvation Army and they give you the full treatment, Bible quotations and reminding you of your sins. You know, the 'holier-than-thou' attitude. Some of the girls lose their tempers and then they've really had it."

"What do you mean?"

"You know, the housekeeper picks on their work, makes them wait longer in line for their linens, gives them bad marks with the staff. Just don't act proud in front of them and you'll be all right." Karen smiled. "You look scared," she said. "It'll be O.K." We passed the foyer where I'd waited for Karen and Lucy earlier today. It seemed a small eternity ago.

The library is next to the large TV lounge. The big lounge, with comfortable couches and chairs and a soft carpet on the floor, was empty.

"No smoking." Karen shrugged her shoulders. "Most of us here are kind of on edge—you know we're not exactly on vacation—and we smoke more than we should, I'm sure. But we've got to have something to calm our nerves." She opened the door to the library—soft green carpet, deep chairs, walls lined with books. Mikki, the Japanese girl, was the only one there; she looked up from a stack of what looked like textbooks and smiled at us.

"How's the tour?"

"Fine." I wanted to stay and talk to her, but Karen pulled on my arm. "Half an hour till bedtime—see you later, Mikki." The Japanese girl waved a slender hand at us; her eyes were already on the book she was reading.

"You don't want to see the dorms down here, do you?" Karen was heading for the stairs. I shook my head.

"The gray dorm is exactly like the ones upstairs and the little one—well, it's just a room with five beds."

"What about Mikki?" I said. "What's she reading?"

"It isn't a secret," Karen said. "She's a medical student. She's got two more years to go before she's a doctor. Being here has made her decide to specialize in obstetrics, so she's got a way to go yet."

"What about her baby?" Somehow I knew the answer. Karen didn't look at me.

"She's arranged for a private adoption. A Japanese lawyer and his wife."

"Oh." There was nothing more to say. Karen walked quickly ahead of me. We had talked about the sort of things that I understand are taboo in here, and maybe I shouldn't have asked. But I didn't think Mikki would mind if she knew. I don't think she'd mind even if I talked to her about it.

"Second floor," said Karen. She was breathing heavily after the climb. "Our home, sweet home." The girls from our dorm and from the pink dorm across the hall were flocking to the bathroom in pajamas and robes carrying toothbrushes and towels. Every time the doors swung open we heard the sounds of water splashing and voices echoing.

"The hospital is at the other end of our floor," said Karen. "You just walk right down there and report to the nurse on duty when your time comes. It beats having a baby in a taxi." She laughed. "Married mothers have all sorts of worries we don't even have to think about, like packing a toothbrush." I could see down the long hall with the shining linoleum floor; at the other end was a door painted baby blue.

"That's the nursery," said Karen. "We've got two labor rooms, a delivery room, nurse's office, examination room and two mothers' wards. A big one for mothers who don't keep their babies, a little one for the ones who do. Isolation is in that end of the house, too—for girls who are sick or on special diets. I guess that ends the tour—except for the schoolroom. I almost forgot." She opened the door halfway down the hall, and I stared at the rows of desks, the blackboard and the maps on the walls.

"Eighteen girls attend high school here," Karen explained. "They get credit for the courses they take, and when they leave they can go back to their old school—they haven't lost any time." The idea somehow chilled me. I thought of little mothers-to-be studying history and math—and when they leave they've got to go back to studying history and forget that they are mothers.

"We don't have any classes for junior high-ers yet," Karen said. "We've got five girls in that age group here now, and the way things are going it looks as if they'll need a classroom and a teacher before many seasons have gone by."

What a day this has been! I guess I won't be so confused when I've been here for a while. By the time March comes around I'll probably know it so well I won't ever be able to forget.

Sunday the 20th

After lights out last night we had a cookies-and-Kool-Aid party on our dorm. One of the girls has an aunt living down the street who brought us the cookies. We sat on pillows on the floor around the goodies, while the pale moon shone in through the windows on our bulky figures huddling together. We were whispering and trying to suppress our giggles. The Major was on duty last night, and our lookout reported that she was safely out of hearing distance in the diet kitchen, having coffee with the night nurse. The party ended in a pillow fight. I crawled into bed early and watched the spectacle from a safe distance. At last only Lucy and Mikki were left on the floor, chasing each other around the island of lockers and chests of drawers. All you could see of Lucy was her bulky white nightgown—her chocolate-brown skin blended with the darkness. Their shadows danced in the moonlight, their movements grotesquely exaggerated. I saw them sit together on Mikki's bed whispering quietly. Then Lucy climbed into her own bed and all was quiet.

I stayed awake and to me the night was full of new noises. The breathing of fourteen girls came from all around me. Someone in the far corner snored lightly and I heard the sound of steps—dragging slippers—in the hall and then the banging of the bathroom door. It wasn't midnight yet and through the open windows came the far-off sounds of the city. Who goes to bed this early on Saturday night?

When I lay on my back and felt my stomach there was a hard bump where it used to be hollow. I'm a big-boned girl and I used to be proud of my flat stomach. Now that thing is growing in there and I can't stop it. All of us in here have that thing inside and we're here to wait till it comes out. We laugh and talk and have pillow fights, and all the time it is growing inside us. They —we—don't talk about what it's going to be like to be mothers, what it's going to be like to give birth to a new little human being. We don't talk about what he or she will look like or if it will be a boy or a girl. I tossed in bed and thoughts going round and round kept me from sleeping.

"Why did You do it, God?" I wanted to cry it out loud. "Why did You let me do it?" But I don't believe in a fate sealed by anything or anyone, I thought. I've got to believe we are free to choose our own destinies. But if I believe that, then I must believe that I chose to be pregnant—chose to come here? I remembered Dorothy's voice: "You want this baby for reasons of your own. . . ."

Around me was the deep even breathing of girls sleeping and I could see their bulky shapes under the blankets. Suddenly I thought I knew what Dorothy had been talking about. Inside me is a baby waiting to be born—the baby didn't choose me anymore than I chose my mother. I am responsible for that baby's start in life, and till now all I've been thinking about has been me, how *I* feel about it, how *my* life is changed. I was glad for the darkness around me. I could feel myself blushing.

A sudden movement on the other side of the room startled me. I saw Mikki sit up in bed, toss her blanket aside, and reach over to shake Lucy. They whispered together, too low for me to hear the words. Then they tiptoed out the door. They were holding hands, I noticed. I turned my head and saw Karen looking my way with wide-awake eyes.

"Mikki must be in labor," she whispered. "They are probably going downstairs to count the minutes between pains and smoke till it's time to go to the nurse. The girls do that a lot. It beats being alone in the labor room."

I looked through the window at the sky and noticed that the Big Dipper had moved into full view from my window, and I wondered if I'd still be awake when it had disappeared behind the tree. That's the last thought I can remember. The next thing I knew, girls were jumping all over the place getting ready for chapel, and my first full day at Booth was under way!

Sunday afternoon

When we got out of chapel the word came; Mikki had a little boy. Her labor had been easy for a first baby, they said.

The Captain led the service. She is a rotund little woman, with a girlish voice and bright blue eyes. Her text for the day was

Romans 6:23: "For the wages of sin is death; but the gift of God is eternal life through Jesus Christ our Lord."

"What a glorious promise, girls!" said the Captain. "Confess your sins and ask Christ into your hearts and eternal life in Him is yours." The Captain's baby-blue eyes smiled kindly at her pregnant congregation. I moved uneasily on the slick, hard, wooden bench of the pew. I wondered if she meant by confessing my sins that I must recognize what I've done wrong, and then I shall find the strength to walk the narrow path of righteousness. But what did I do wrong—and where goes that right path?

Sixteen-year-old Cora from our dorm stood next to the altar, her long blond hair framing her young face, her hands clasped over the protruding stomach. Her voice rose clear and rich: "My faith looks up to Thee, Thou Lamb of Calvary, Saviour Divine." I could close my eyes and forget about the words. Worship in music is something I can understand. I felt at peace for a little while.

Monday the 21st

Miss O'Connor has been assigned as my case worker, or I've been assigned to her, depending on how you look at it. She is quite young, maybe twenty-five, plumpish, blond. Mostly I noticed that she isn't pregnant! Her office is a tiny cubicle, almost like a cell. On the wall are two watercolors of Paris in the spring. Paris —Mother and Dad—I'd love to go there one day. I'd love to walk along the banks of the Seine in spring and sip wine at a sidewalk cafe and be in love. I wouldn't have a care in the world and my skirts wouldn't be getting too tight around my waist.

Miss O'Connor offered me a cigarette (I took it) and said she would help me fill out the standard form: my name, the baby's father's name, record of illnesses in both families, anything that might be important about the baby. I gave my phony name and said I didn't want to tell the father's name.

"That's all right," Miss O'Connor said. "Just tell me his age, race and occupation."

"He's thirty-five, white and in business for himself," I said, and the thought of Gene didn't seem to have any connection with the baby inside me.

"You may apply for aid from the state," said Miss O'Connor. "They will pay for your stay here and your delivery. They also give you eleven dollars a month in pocket money." A ward of the state, I thought. I got myself pregnant and somebody's tax money is going to pay for it!

"Oh, no!" I said. "Isn't there another way?"

"Of course," Miss O'Connor said, reassuringly. "The cost of your stay here is a hundred dollars a month, plus seventy-five dollars for your delivery and postnatal medical care for you and the baby. You can pay that yourself."

"That'll be six hundred and seventy-five dollars," I said. "I don't have that kind of money."

"You may pay it in installments whenever you can afford it," said Miss O'Connor. "We trust you to keep your word and pay us back when you can—and we don't give you any time limit."

"You mean I don't sign something and promise a certain sum every month?" It was hard to believe. Miss O'Connor shook her head. "The Salvation Army doesn't operate on a profit basis," she said. "They believe that they work for Christ in looking after his children—you. They ask you to pay later—only so that they can help others like you."

"You mean they've got that much faith in us?" I said. "In unwed mothers who don't even tell them their names?"

Miss O'Connor smiled. "I've been here for three years and have never heard of a girl who hasn't paid her debt. Several girls pay more, just to show their gratitude." I sat back in my chair and took another of her cigarettes. What I'd just heard didn't sound like the good old U.S.A. in the mid-twentieth century.

"But that's still an awful lot of money," I said. "I may not be able to pay it back even if I want to. I'm going to support myself and the baby, you know." I thought she would say something about the advantages of adoption, but she just looked at me and said, "The decision is yours. I'm sure you'll choose what is best—for both of you." I don't know why everybody keeps saying they know I'll make the best decision. I sure as hell don't know that myself. It seems as if whatever I choose will be wrong. I can't imagine giving up my own baby, and I can't imagine how that baby can get along with me as a mother.

"There is one way you can cut your expenses while you're waiting for the baby," Miss O'Connor said, interrupting my hopeless train of thought.

"You can go to a wage home and stay here only for your last month."

"What's a wage home?" The thought of leaving this safe place scared me.

"They are family homes nearby where a girl gets room and board and ten dollars a week for doing light housekeeping and baby-sitting. You still 'belong' here, so to speak; you come in for your medical checkups and get your supply of iron, calcium and vitamin pills. You may take along and wear the maternity clothes you need. And of course you may come back here to stay any time you want to."

"Then I'd only owe you one hundred and seventy-five or maybe two hundred dollars by the time I leave," I said, after doing some arithmetic. "That sounds a lot better—but I hate to go outside and be an unwed mother in hiding again. In here I'm just one of the girls."

"You'll get all the privacy you want in a wage home," said

Miss O'Connor. "We screen the families carefully before sending any girls to them."

"I'll give it a try," I said. Miss O'Connor will call a wage-home mother and have her come here to meet me first before I decide to go there. I'll probably stay here for a week anyway, for medical checkups and all that.

Before I left her office, Miss O'Connor told me that if I change my mind about keeping the baby she'll be glad to make arrangements with an adoption agency. The Home works both with a state and a local agency. She said they would do everything to find exactly the kind of family I would want for the baby. I shook my head and said, "No, thanks," and she smiled.

"Drop in whenever you feel like talking, Jean." I said thank you, and I will. She's really O.K. Easy to talk to and no pressure.

I looked at those watercolors on her wall and Paris seemed an awfully long way off. I'm twenty-one now and I'll be thirty-nine before this baby is eighteen. Am I going to be alone with the responsibility for the two of us till then?

Monday night

I'm beat. What a day. After I talked to Miss O'Connor I went upstairs to take a nap. The nights are kind of noisy and I haven't gotten used to it yet, so I don't sleep much. The dorm was empty. Monday morning everybody is either working or going to school. No sooner had I flopped on my bed than my name was called over the intercom. I jumped three feet. It was the first time I'd heard my new name over that thing and it shook me. I was called twice before I had sense enough to answer "Coming."

Downstairs in the office the skinny, tall Major with black hair pinned in a severe bun was waiting for me. Karen had pointed her out earlier and told me her name is Laski and that she came to this country from Poland in 1937. She still speaks with a thick accent and the girls imitate her.

"What were you doing in your dorm, Jean?" she said in a stern voice.

"Taking a nap," I said, and watched how the sudden smile started in her dark eyes and made her face look soft and young.

"No girls are allowed in the dorms during working hours unless they are sick," she said slowly, but still smiling. "I guess you need a job."

"I'm going to a wage home next week," I said, hoping to get out of the working bit.

"To be idle is not good," said the Major. "You will work in the kitchen, set the table for the housekeeping staff. It is an easy job."

"Thank you," I said.

"Are you comfortable in your dorm? Have you met friends?"

"Yes, thank you." It was hard to stand still under the Major's kind eyes.

"I am the housemother," she explained, "but you work for the cook. She will tell you what to do."

"Thank you," I said again. "But what do I do with my suitcase? It's still in my dorm."

"Take it downstairs. The housekeeper will open the storeroom for you." I said thank you and ran upstairs. There in the dorm was the housekeeper standing by my bed holding my empty suitcase. She looked pretty mad.

"Don't you know you can't have personal luggage in the dorm?" she said. "What do you think would happen if all of you kept suitcases under your beds?"

"I'm sorry," I said, trying to make my voice sound like I meant it. "But I'm new. I haven't had time to take it downstairs—I was going there now."

"So let's go." She tossed the suitcase at me and I grabbed it before it fell. Mrs. Larsen marched ahead of me muttering something about "think they're so smart" and I walked behind her, thanking heaven that Major Laski had given me a job in the kitchen instead of in Mrs. Larsen's department. Before she locked my suitcase in the storeroom in the basement, she turned and said, "Sure you ain't got something in here you need before you leave? I got other things to do than run up and down stairs unlocking this room for scatterbrained girls." My blood was boiling, but I remembered Karen's warning. Keep your temper with the housekeeping staff or you're in trouble. I shook my head, forced a smile and said, "No, I'm sure I won't need it." Mrs. Larsen looked me up and down.

"You don't look pregnant now—but you'll need smocks later. Might as well pick 'em out now. Save me a trip." She waved toward the rack of maternity tops and skirts lining one wall in the storage room. "What's your size? Ten? You'll get bigger than that. Take twelves!"

The smocks were mostly bright cotton prints. I took a brown and a black skirt and three tops with big pockets. They didn't look too washed-out. I remembered Karen had a green two-piece corduroy maternity dress. If it belongs to the Home maybe I can get it when she delivers. Mrs. Larsen left me in the basement corridor with a friendly "Take your clothes to the dorm and don't waste any time about it. We don't allow girls in the dorms during working hours!" Thank goodness *she* didn't catch me taking a nap.

I got the clothes put away in my locker and found Karen in the smoking room downstairs. Her job is in the scrub room next to the kitchen, where they wash the big pots and pans. I told her about my run-in with Mrs. Larsen. "I hope the cook is better," I said. Karen smiled and shook her head.

"They're bosom buddies," she said. "The cook and the assistant cook are sisters, Mrs. Tebbits and Mrs. Coralis. Mrs. Tebbits is the youngest and she's the head cook. Mrs. Coralis is the sourest. When the three of them aren't working or praying, they're scolding us or gossiping about us."

"Is it all that bad?" I was getting worried.

"Nah," said Karen. "They don't really hurt us, and deep down they really mean well—I hope." She pulled on her cigarette. "You know they think we're really sinners and that we ought to repent and get punished. It gripes them when the Salvation Army officers aren't more strict with us."

I said, "How do you put up with it?"

"They're really doing an awful lot of good for us," said Karen. "Where would we go if we couldn't be here? So maybe they want us to eat humble pie and lick their boots and say 'Yes, Ma'am, I'm sorry, Ma'am!' You can't really blame them."

"Why not?" I said. "Can you talk to the officers about it?"

Karen shook her head. "No, they're more like supervisors and administrators. They don't work directly with us as much as the housekeeping staff does." She got up and stubbed out her cigarette. "Time for kitchen duty. Let's go."

Karen introduced me to Mrs. Tebbits, the head cook, who smiled and said she was glad to meet me. Mrs. Tebbits has false upper teeth, and they fall with a soft click against her lower plate every time she talks. It sort of distracts you from what she's saying. She didn't sound too bad to me. She was very friendly when she showed me where to get the plates and the silverware for the table.

Only six of them eat at "my" table in the kitchen. The two cooks, the housekeeper, the woman in charge of the laundry, the janitor and the gardener. The cooks take turns with the Lieutenant in presiding at the table in the girls' dining room. My job is to set the table and put the food out before I go to eat with the others. Then after the meal I clear off the table again and help the girl who runs the dishwasher. Once a week I am supposed to help wash down the walls in the kitchen and the pantry. In between mealtimes I can do as I please, except, of course, stay in the dorm. I can watch TV, use the library, work with ceramics or paint, or sew in the craft room or walk in the park. In the afternoon I can get one two-hour pass every day, and I can get an eight-hour pass once a week. Of course, if I break any rules—get caught in the dorm during the day, come home late from a pass, or conduct myself badly in the eyes of the staff—I can lose my pass privileges. I just hope I can keep my temper.

The girl who runs the dishwasher is thirty-five years old. Her name is Maggie and she told me she wants to keep her baby.

"I've got enough money saved and my boss promised to keep my job open for me. But, you know, it's hard on a kid not to have a father." She pushed the hair back from her forehead with a wet hand.

"I just don't know if I've got the right to keep my baby," she said in a low, weary voice, and you could tell she had thought about it over and over again. The girls don't talk much about their reasons for giving up or keeping their babies, and it's an unwritten rule never to say what you think about another girl's decision. We go through this pregnancy together—but each of us faces the future alone.

Monday afternoon is weight-time. We all line up wearing our lightest smocks. We're called in alphabetical order; some of us have to wait in line for over an hour for our turn. Karen, Lucy and I have last initials S. and T., and Lucy fumed about it. "I'd have picked myself a name starting with an A if I'd known," she said. "That 'S' isn't my name, anyway."

Some of the girls are compulsive candy eaters and starve themselves all day Monday to get by Nurse Simpson's sharp eye. Nurse Simpson is tall and white-haired and strict, but her blue eyes are

SEPTEMBER

warm. When she restricts an overweight girl on passes, or sends her upstairs to sleep in Isolation or to eat in the diet kitchen, it is to help her.

"You want an easy delivery, don't you? And you want to look as if you haven't had a baby when you leave, don't you?" she says in an icy voice, and the unhappy girl whispers "Yes" and heads for Isolation.

Most of us want to look as if we've never had any babies when we leave here, of course. Some of the girls manage to keep their total weight-gain down to five pounds or less and look fashionably slim when they leave. If we thought we could starve ourselves into *feeling* as if we'd never had a baby, I bet we'd quit eating all that junk between meals. But we know we can't do that, so most of us spend our few extra pennies on cookies and candy and Cokes at the corner drugstore. I already weigh ten pounds over my normal weight and Nurse Simpson said I'll have to watch out.

"You've got six more months to go and only five more pounds allowed," she said.

If I can only get over this restlessness. There are plenty of things I could do here, but most of the time I end up just sitting around doing nothing. Most of the girls seem to feel the same way. All we do is talk about our pregnancy symptoms.

We just had mail call. The Lieutenant's voice announced it over the intercom and we all ran for the television lounge. I haven't told Dorothy I'm here yet so no one can write to me, but I still ran to stand on the outskirts of the eager group of girls pressing around the Lieutenant. Karen told me a lot of the girls get their mail routed through other Salvation Army post-office boxes in other cities.

"You stay out of sight for months, and someone you know is bound to try to look you up," she said. The thought makes me shudder. I hope I've covered my tracks well enough.

Tuesday the 22nd

Tuesday is a confusing day. First we strip our beds and spend at least an hour in line to dump our dirty linen and to get handed our clean linen. Then we rush back to make our beds. Half of us head for the hospital floor for our medical examination while the other half go to their jobs. I had my first checkup today and Doctor Norwad said I am in "perfect physical health." I wish I could say the same for my mental state.

Karen didn't have a checkup today, so she cleaned off the breakfast dishes in the kitchen for me. I helped Maggie rinse the dishes and put them in the dishwasher. We had oatmeal for breakfast.

"We always do, it seems, when we have checkup," said Maggie, as she scraped away at the hardened leftovers. The stuff is almost impossible to get off if you don't do it right away. We had just finished, when it was time to start all over with lunch! We put the hot dishes straight on the table from the dishwasher—not even time for a smoke in between.

Dr. Norwad holds a motherhood class in the TV lounge after lunch. We're all required to attend and can ask any questions we like. He talked about how our bodies change to accommodate the growing fetus and what we can do to keep ourselves in physical shape during the pregnancy. The slim young Lieutenant demonstrated exercises to do to prepare for the delivery.

"What's the best exercise?" asked a girl from the pink dorm.

"Other than a mile's brisk walk daily—this one," said the doctor, and demonstrated by pushing himself away from the table. "Practice whenever you're tempted to overeat." Karen has attended the whole lecture series before. She says the doctors cover the entire development of the baby from conception to birth.

"The last lecture is on contraceptives," she said, and smiled. "I guess they don't want us in here twice."

Wednesday the 23rd

There's a new girl in Mikki's bed in our dorm. She was there when I came back from kitchen duty after lunch. Her bags were still packed and she wore street clothes, a beige cashmere sweater over a brown and beige wool skirt. She didn't look the least pregnant and her face was skinny below straight ash-blond bangs.

"Hi, there," I said and she looked up quickly as if she expected me to be someone who knew her. That scared look left her eyes and she smiled when she saw I was a stranger.

"I'm Jean T. from Chicago, due in March," I said. The new girl nodded slightly and said, "I'm Marian C. from Norfolk, Virginia." I recognized her New England accent and wondered if she was from Boston.

"I'm due in April," she said, and must have noticed my eyes on the letter she was writing. "For my boyfriend," she said, and I wondered if he was the father of her baby. Her gray eyes were serious; she didn't look like someone who would sleep with a guy she didn't love. But in here you never know. When a girl talks about a boyfriend she's going to marry when she leaves, he is never the father of the baby. Like Clara who left just before I came. Karen told me Clara's boyfriend had followed her across the country, and begged her to marry him and let him adopt the baby, but she refused. Her baby went out for adoption through the agency, and Clara married the boy. She said that if she'd kept the baby she would never have known for sure if she married him just so she wouldn't have to give it up.

"We're going to get married—later," Marian said, smiling in spite of a fresh flow of tears.

"And the baby?" The question popped out against my better judgment.

"We're giving it up for adoption," she said and went on rapidly. "He doesn't want it that way either. He wants his baby as much as I do."

"His?" I said, and Marian looked hurt. "Of course!" she said. "You don't think I'd have someone else's baby, do you?" I shook my head and wondered what on earth had brought her here. Major Laski's voice broke in over the intercom: "Marian C., Marian C." The new girl looked as if the name had nothing to do with her. The Major called again: "Marian C., Marian C."

"Coming!" several girls yelled in chorus and I nudged Marian.

"You're wanted in the office, they're calling your name." She still looked dazed when she walked out the door.

"She's sweet," said Lucy. "Wonder how she got into this mess." Cora, the girl who sang in chapel last Sunday, looked up from the confession magazine she was reading and laughed out loud. "Just like the rest of us, sweetie—don't you know how babies are made?"

Several girls laughed, and Cora turned with a toss of her bleached blond hair to Lisa on the next bed.

"You know how babies are made, don't you, honey?"

The dorm was quiet, and I knew why Karen had called Cora "a bad one." Lisa is fourteen and seems to be mentally retarded. She was brought here from Juvenile Hall and no one knows for sure when her baby is due. Lisa's pale gray eyes blinked behind the thick lenses of her glasses and she stammered, "I . . . I don't know. I didn't do nothing wrong."

"You kissed a boy, didn't you, honey?" Cora's voice was mockingly sweet. Lisa nodded. "Didn't he tell you that's how babies are made?" Cora feigned surprise, and Lisa looked around helplessly.

"N-no!" she whispered. Cora jumped off her bed and put her arm around Lisa's shoulders. She was enjoying herself.

"But you know there's a real live baby inside your tummy, don't you?"

Lisa's smile was sudden and glowing. She nodded happily. "The doctor told me," she said proudly. "Pretty soon the baby will come out and I'll play with it." The smile left Lisa's face as suddenly as it had come and she looked intently at Cora.

"You're nice to me," she said. "How does the baby come out of my tummy?" She put her hand between her legs. "Does it come out here?"

Cora nodded. "It sure does, honey," she said. "Same place it came in." Lisa pulled away from Cora's grasp.

"You're just kidding me—you always do."

"That's enough, Cora," said Lucy, her voice sharp. She walked to Lisa's bed and took her hand gently. "No one will hurt you," she said softly. "The doctor will help the baby come out of your tummy while you're asleep." Lisa beamed at Lucy.

"You're nice. Will you play checkers with me?"

"Sure, Lisa. Let's play in the lounge." The door closed after them and Cora threw herself backward on Lisa's bed.

"God Almighty, that kid gives me the creeps," she said. "You're nice to her and she hangs on like a leech."

"Don't be nice then," one of the girls said.

"Leave Lisa alone," said Karen. "She doesn't hurt anyone."

"Blah, blah, blah." Cora grimaced. "She isn't very smart, her parents don't want her and poor, poor Lisa." She looked around at all of us. "You all make me want to puke. So who do you think you're kidding? We're all alike in here, and don't you try to say 'Poor Cora' or I'll beat the shit out of you. I never had any parents and I worked all my goddamned life. I was dumb enough to get into this mess, and it isn't gonna happen again. Lisa isn't smart

enough to hit back and you can't save her. But I've learned *my*
lesson. I know damn well nobody's gonna love me and pay for
my living just because I'm nice and sweet. Everybody's too busy
grabbing for themselves. And let me tell you something. Old
Cora's gonna get right in there grabbing with all she's got." No
one tried to answer, and Cora laughed to herself as she curled
up on her bed with the magazine again.

So that's our little girl who sings like an angel!

Thursday the 24th

I'm leaving for a wage home this afternoon. Miss O'Connor
called me to her office to have me meet the wage-home mother.
Mrs. Duncan is young, pretty and expecting her third baby in
April. She was on her way to a luncheon and didn't have much
time to talk. That's just as well. I don't want to get to know her
and I don't want her to know me. The Duncans live in the hills
outside town and Mrs. Duncan said she hoped I didn't mind being far from neighbors.

My feelings are so mixed up. I don't want to meet people on
the outside—but I'm scared of being all alone with my thoughts.

Later Thursday

I sneaked down to the corner and bought a cream puff for
Mikki. Lucy told me she liked them and I wanted an excuse to
go visit her in the mothers' ward. Her bed was by the window
where she could see the park and the old Russian gardener caring
for his flowers all day long. On her bedstand were two books,
the Bible and a thin little volume I recognized, *The Prophet* by
Kahlil Gibran. Mikki was looking out the window, her pale, slim

hands folded over her flat stomach where there wasn't a baby anymore. I thought maybe she wanted to be left alone, but it was too late. She turned her head and smiled when she saw me.

"Hi, there, Jean. How nice of you to come."

I held out the white bakery bag. "I brought you a cream puff."

"I love them!" Her eyes sparkled when she smiled, and I smiled back. "So do I."

"Let's have a party then." Mikki sat up in bed and divided the cream puff carefully, brushing one half against the other to share the cream.

"Have some," she said, offering me the top half. She looked like an impish little girl sitting there cross-legged on the bed, licking powdered sugar off her fingers. I kept wishing that was all we were: two girls sharing a cream puff and laughing about it, with the sun shining in on us through the tall windows. But Mikki had gone the long road from the decision to give away her baby to this moment, when she was actually facing it. My road was ahead and I was scared. I sensed in her a strength and an awareness I know are lacking in me.

"You're going back to school now, aren't you?" I hoped she wouldn't mind talking about it. Mikki nodded. "I sure am. It's going to be great working for what I want after all this fooling around."

"You mean having a baby and all?"

"Oh, no!" Mikki's dark eyes were suddenly serious. "This was no fooling!" She lay back on her pillow and looked up at the ceiling. "I don't think I ever did anything real before this. I was doing pretty well in school and some people used to think I was bright for my age." She laughed, a funny little laugh. "What a mess my mind was. I was terribly self-centered, but I had a hard time finding a real me. People—I mean my Dad and my teachers—used to tell me I would turn into something great one day, that I had it in me. Maybe I even believed it myself sometimes, but I was really scared that I wouldn't measure up." Mikki turned her head and looked straight at me.

"And so I got pregnant." She smiled but her eyes were serious. "Hope you don't mind me talking like this, Jean. But we won't meet again and when I leave here I won't speak of these things again."

"I don't mind," I said. "I really came because I hoped you'd talk. . . . I wanted to ask you something."

"What?" said Mikki. "I'll answer if I can."

"Why are you giving up the baby?" The question hung between us. Mikki was sitting up in bed, her hands calm in her lap, and behind her I could see the trees in the park lifting their gold and red crowns toward the September sky.

"The baby was never mine to give up," Mikki said, slowly. "I bore him and he came through me, but he doesn't belong to me." She took the little book from her bedstand and leafed through the pages till she found what she was looking for.

"This is what *The Prophet* says," she said, "that our children are not our children but part of life's longing for itself. That they are as arrows and we, the parents, the bow. God is the archer." I knew the chapter she was talking about. "I *did* want to keep the baby at first," she said. "I thought it was because I wanted to show the world that a child and his natural mother belong together, no matter what. Maybe I have grown up a little since then, but not enough to help a child meet the world. I wish I could try—believe me." Mikki's voice was low. "By parting now the suffering is mine alone. If I keep him, both of us may suffer."

"Do your parents know?"

"I told mother my baby was due in November. Now she thinks it was stillborn. We haven't told my father; he has been ill. When I have my degree I may tell him. I don't know yet."

I had just one more question I wanted to ask: "Why did you choose a private adoption?"

Mikki laughed a little. "I *am* Japanese," she said. "It might not be so easy for the agency to place my baby right away. I wanted to be sure he had a good home. I went to a Japanese lawyer and, as it happened, he and his wife were looking for a baby to adopt. I was lucky."

"I wish I knew what is the right thing to do!" I said. Mikki nodded.

"I know what you mean, but there are no guarantees. You must be strong enough to act without being sure that what you do is absolutely right."

"I came to say good-bye," I said, "and thank you. I am going to a wage home this afternoon." Mikki reached out and took my hand.

"Whatever you do, Jean, make up your mind now," she said. "You have all the facts and you need time to strengthen your conviction. What you decide will be with you for the rest of your life. Your reasons must be strong enough and clear enough to withstand doubt." Mikki smiled and squeezed my hand. "I didn't mean to lecture you. But I know what you are facing. I even wrote my thoughts down. I could look them up when I got emotional."

"Thanks," I said. "I'll try to remember—and good luck!"

"Good luck to you." I saw her open *The Prophet* again and settle back against the pillow, but her eyes were on the old Russian gardener below the window, watering his flowers.

Thursday evening

I'm waiting for Mr. Duncan to pick me up. I had already packed my things when Mrs. Duncan called and said her husband would be late. I didn't feel like eating supper and came downstairs to the smoking room instead. Marian was in a corner staring at the dirty concrete wall and crying without a sound. She didn't turn her head when I came in.

"Chow time," I said. "Aren't you going to eat?" She shook her head without looking away from that wall. Maybe she was sick.

"Anything I can do?" I said and she turned to me, her eyes swollen from crying for a long time.

"My boyfriend just called," she said. "He wants me to come back and marry him. He's borrowed money for my ticket."

"So what's the problem?" I said.

"Mother doesn't want us to get married. She says people will talk, if we have a baby five months after the wedding." Marian saw the look on my face and blushed. "Mother loves me," she said quickly. "She doesn't want me to get hurt." I kept thinking it was none of my business—it was Marian's problem. I had my own to think about.

"Known your boyfriend long?" I said, and the glow came back in Marian's eyes the way it always does when she talks about him.

"Ever since I was a little girl," she said. "He's in graduate school now. He'll have his master's degree in chemistry in two years. Mother says I can marry him then if I still want to." Her voice suddenly lost its eagerness and she stared at me darkly. "That is, if Chris still wants me after I've given away our baby."

"Have a cigarette," I offered, but she shook her head. I lit one for myself, and noticed that my hands were shaking quite a bit.

"What should I do, Jean?" Marian looked at me but I knew she was talking to herself. My opinion wouldn't matter.

"Since Daddy died Mother doesn't have anybody else," she said. "But the baby—Chris and I will be parents together whether we keep it or not—we're responsible, aren't we?" I puffed on my cigarette and nodded; Marian was looking at me intently.

"We are responsible for the child first, aren't we? Say something, Jean. Agree with me—I need someone to agree." I wanted to tell her yes, but I shook my head instead.

"What difference does it make," I said, "if anyone agrees?"

Her eyes went blank for a second. Then she suddenly jumped up.

"Got a dime?" she asked, her hand already stretched out. "I'm going to call Chris collect and tell him I'm coming home. I've spent half my life waiting around for other people to make up my mind for me. Mother, Chris, anybody who'd listen. By God—" she smiled a little—"what difference does it make? I've got to do what's right for me whether someone else agrees or not." She took the dime I found in my pocket and was gone. I heard her run down the hall and slam the phone-booth door shut after her.

So there is one of us finding her way out.

At the Duncans', Saturday the 26th

It is nearly midnight and I don't have to go to bed. Over the hill hangs a full moon; I have turned off the electric light and lighted a red candle on my table.

My room is small, with a bed, a chest of drawers, a table and a chair. I have room to walk one step in one direction and two steps in the other, that's all. But it is all mine. I thought I would feel lonely and the white walls would seem too close. Instead, I feel a peace inside that is an unexpected contrast to the restlessness I've always felt before when I was alone. But then, of course, I am not alone here. There is a little somebody inside me and I am beginning to get used to the thought.

I have hung two pictures on my wall. One is a woodcut of three boats sailing into the sunset, done in brown, rust and black on unbleached coarse sailcloth. I bought it in a small gallery in Greenwich Village a few years ago. The other is a reproduction of a painting by Diego Rivera that I tore out of the art section of *Time* magazine. It is called "Mexican Child" and is of a small girl sitting with her hands in her lap and staring ahead with large, solemn brown eyes. I love her eyes. It is as if she says: I am not afraid—you will care for me.

Most of my books are stored in Dorothy's attic, but some I couldn't bear to leave behind. They stayed in my suitcase in the locker at the Home but in this little room they are with me. A. A. Milne's *Winnie-the-Pooh,* Kenneth Grahame's *The Wind in the Willows,* Antoine de Saint-Exupéry's *The Little Prince,* and of course, *The Prophet.* They are all loyal old friends.

So this is my little kingdom; my room with a view.

Monday the 28th

The house rests in the valley surrounded by rolling hills dotted with clusters of fine old trees. The blacktop road leading to the city goes along the rim of a hill high above us. A dirt road winds its way down the hill to our house and then on past a grove of trees and around a bend to our nearest neighbor, half a mile away.

The traffic on the blacktop road is light. The mailman comes once a day in his red-white-and-blue truck and leaves mail in the two boxes by our driveway. Three times a week the milk truck and the bakery truck come down in a cloud of yellow dust.

We are three miles from the highway where the buses go to town, and fifteen miles from the city itself. I get a day off once a week. Mr. Duncan will take me to town when he leaves for work in the morning and bring me back when he comes home in the evening. But I have nowhere to go, so I will only leave here once every two or three weeks when I go to the Home for my medical checkup.

The house is only a year old, quite large, with seven bedrooms, four baths and three fireplaces, no less! Four of the bedrooms, with two baths and a playroom, are in the children's wing, and that's really all I have to worry about. Josephine, the colored maid, comes every Tuesday and Friday, sometimes on Saturday, to do all heavy cleaning and ironing. I just look after the children, straighten up their rooms and feed them whenever Mrs. Duncan is away.

Kids have always made me feel awkward; I never knew what to say to them. I've never even been a baby-sitter. But Peter and Lori are nice. They don't ask for a lot of attention and they do as I say. Peter is eight and in the third grade. He is in school most of the day and when he is home he's got all kinds of projects going. His room is full of things I always thought little boys ought to have. Like rock collections and an abandoned bird's nest, an old steering wheel, a homemade slingshot, and Alexander the turtle in a cardboard box.

Lori is only four and plays alone most of the day. She has brown eyes below blond bangs and lives quite often in a world of pretend. I hear her talk to her make-believe friends and pets; sometimes she sings to them. Maybe it's because there are no other four-year-olds in the valley.

The new baby is already part of the family. Mrs. Duncan is knitting a little white baby-sweater, and Lori rocks her teddy bear to sleep and sings about the baby. Peter works in spurts on his project; he is building a car for his new brother (Peter is sure it will be a boy), and that's what the steering wheel is for.

My presence doesn't seem to make much difference to the household. When Mrs. Duncan is home, I do my work quickly and go to my room or for a walk in the hills. Sometimes in the evening I am invited to join them watching television, but I enjoy the solitude of my room so much I'd rather stay here. Also, I don't want to be with the Duncans more than I have to. I don't want to be tempted to talk about myself. Watching the family doing things together hurts a little bit, too. I am just plain envious, I think, and it makes me feel more restless and alone.

Tuesday the 29th
Late evening

Mikki's words, "Make your decision early," have been bothering me every day. Last night I wrote Dorothy and told her to look around for an adoptive family, just in case. I haven't made up my mind yet, but if I am going to part with this baby I must start making plans now. Just because it's growing in me doesn't mean it belongs to me. I know that. It would be like saying that a flower belongs to the soil.

It is strange how the love between mother and child develops from one-ness to separateness—independence. And the love between man and woman develops the other way.

There is the daylight coming over the hill. My candle is burning low and dripping. When I blow it out, the window frames a square of light gray pre-dawn.

OCTOBER

Monday the 5th

I put Lori down for her after-lunch nap and ran barefoot up the steep dirt road. I still don't look pregnant when I wear a loose shirt over my dungarees. Running uphill takes my breath away. I'll probably have to give that up soon.

The view from the rim road is beautiful. The grass on the rounded hills has dried to a rich yellow, and the clusters of trees range in color from a deep green to orange, yellow and rusty brown. Yellow and red wildflowers grow by the wayside. I picked enough for a bowl in my room.

I saw the mail truck coming and stopped to watch it. The heavy beating of my heart wasn't from running. It was about time for a letter from Dorothy. The mailman smiled and waved at me.

"Are you Jean Thompson? Staying with the Duncans?"

I nodded and he handed me the bundle of mail, several magazines and letters, and on top an airmail envelope addressed to me in Dorothy's strong, slanting handwriting.

"Just want to make sure when there's a new name that somebody didn't get the wrong address." The mailman grinned and waved again as he took off down the road.

I walked back slowly, feeling the gravel under my bare feet. I looked at that envelope addressed to Jean Thompson; it seemed strange it should have anything to do with me. I ate my lunch before opening it, trying to let my mind go blank. Letting the thoughts wait makes me feel as if I have a choice. Dorothy wrote:

I am proud of you and think you are wise in making the arrangements for the baby's adoption this soon.

A good friend of mine, who lives in the city where you are, knows a young couple there who adopted a baby three years ago. They would like to have another child now. My friend says they are fine people. He is a high school teacher. If you are interested, I will let them know about you and their lawyer will get in touch with you.

I miss you, and you've had two phone calls. I told them you were in New York for a couple of days. What are you going to tell your family about Christmas plans?

Let me know if there is anything I can do for you.

Why don't you want an agency adoption?

Love,
Dorothy

How typical of her. She is matter-of-fact and to the point, outspoken and a stickler for details. I don't know what I'd do without her. Now, why couldn't Mother be like that? Blow her stack when I do something she disapproves of but accept me for what I am. Oh, I know she means well, but we just don't talk the same language. She'd die of shame if she knew what I am doing. She belongs to the Salvation Army Auxiliary back home and I've helped her wrap Christmas gifts for those "poor misguided, unfortunate girls." She never believed it could happen to the daughter of anyone she knew, especially her own! In her mind those girls were "different."

The couple Dorothy wrote about—are they the right family for my baby? How can I be sure? I wish the baby were here to talk to. I wish I didn't have to decide alone what is best for both of us.

Tuesday the 6th

The letter I've written Dorothy is right here, sealed and stamped. I can still tear it up.

I am telling her to write that young couple. To tell them I want to see their lawyer as soon as possible—and then I want to meet them. Maybe it will hurt more that way, but how else can I be sure they are the right family for my baby? I've got to make sure. Am I doing the right thing? Am I going against a natural law that says a child belongs with its natural mother?

In my letter to Dorothy I say I am sure. From now on I must act as if I am sure—even if I can never be. When I mail the letter, I have committed myself. From then on my fears and my doubts must be mine alone.

Must it be this final? Dear God, what have I just done?

Later . . .

I ran all the way up to the mailbox and flipped up the little red flag on the side, so the mailman would be sure to pick up my letter. I could see his red-white-and-blue truck up the road.

Then I came back to my room and tried to sit in the chair by the window, my hands touching the white painted window sill. The afternoon sun was slanting through the trees along the edge of the hill. I moved my fingers over the smooth surface of that sill, trying to put my whole being into touching the wood, trying to stop the thoughts, the fear. I closed my eyes and tried a desperate wish—let me get out of here, out of this waiting for something I can't anticipate—except that it will hurt. I try to fill the time with simple surface thoughts and, for a little while, I can almost make believe that is all there is.

And then something goes wrong inside. I am afraid this time it will break and carry me away out there to drown in the oily smooth blackness. I mustn't scream. I mustn't let go! God! I say it out loud and my voice is calm, still under control.

Give me anything but this quiet waiting, this stillness and loneliness. People—but I don't want people. I'm sick of them, the shells that are people and the empty voices coming out of them.

Mrs. Duncan knocked on my door and peeked in. She looked concerned.

"Jean, are you all right?"

I made myself smile and nod; my throat was too choked up to talk. Mrs. Duncan was pretty in her new maternity dress.

"We're going out for supper—give the children hamburgers and don't let them watch too much TV before bedtime."

Later . . .

I can't let myself go like that again. I'm walking a tightrope and I've got to keep my balance, so I've taken a stand. That's that. Alone, I could afford to wallow in self-pity, but others are involved and I'm committed to them. Help me hold off my emotions just a few more months—till I've fulfilled my commitment to the baby and the adoptive family. I'm on my own after that and I can only hurt myself.

These months are the real thing. I can't pretend otherwise. I was always a dreamer. I thought my dreams were real and my actions not so important. Dreams are easy; they never leave scars.

You're never stuck with consequences of your bad dreams. Just dream another one.

Wednesday the 7th

A note from Dorothy—enclosed was a letter from Patrick. Dorothy wrote:

> I'm enclosing a letter for Jean Thompson from someone I've never heard of—how come he knows that name and not your whereabouts?
>
> Surprise! My thoughtful husband thinks I should go to California to visit Aunt Helen before Christmas. Wouldn't it be grand if I can come to see you? I'll let you know more about it later. . . .
>
> All my love.

I made myself an extra cup of coffee and curled up on my bed before opening Patrick's letter. He wrote:

> Dear strange girl from the train . . .
>
> I have a terrible problem. You may be the girl of my dreams, and I let you go without trying to find out. Now if I rush back to see you, you may not be my dream girl. Then I won't even have that dream to keep me company.
>
> The red leaf isn't in the envelope by accident. It stuck to my foot when I walked home from class in the rain, an autumn greeting to you. . . . I am coming back east on my Christmas vacation. Will you meet me then, or don't you like my dream?

I wish I could tell him everything. But maybe he wouldn't write to me then. He is on the outside, where people put labels on each other and don't look back of the labels. I can't tell Patrick that I'm going to be an unwed mother soon. What about his dream then? Would he still send autumn leaves?

Saturday the 10th

Karen just phoned; it was really good to hear her voice again. She said she's discovered a small café down the street where they have the best coffee in the world; it's a truck stop, and the guy who owns it makes great hamburgers, too.

"Let's go there Tuesday when you come in for your checkup."

Karen said Marian has gone to a wage home. She'll only be there a month and then go home to get married before Christmas. Cora had her baby and didn't even want to see it, Karen said. She left the home yesterday looking very slim and very striking in a new suit and with a new fluffy hairdo. Karen said she was going to Hollywood to get a job. Lucy has delivered and gone, too. Lisa was moved downstairs to the small dorm early this week. Karen said the other girls had been teasing her.

"And you," I said, "how do you feel?"

"Couldn't be better!" Karen laughed. "I've been having some real sharp dropping pains lately. The more it hurts the sooner it's over!"

Lori came rushing into my room screaming at the top of her voice, "The bird bit me, the bird bit me!" I left the typewriter to take her in my arms.

"Birds don't bite. They don't have teeth," I said.

"This one does," Lori sobbed. "It is a red bird and it has giant black teeth." I said perhaps it was a pretend bird, and she stamped her foot. "That doesn't make any difference. Pretend birds can bite too." She looked at me with tears rolling down smudgy cheeks.

"I know they do, but now you're safe and maybe we can find

a big pretend dog to chase the bird away?" That made her laugh and she ran outside to look for the dog. In Lori's little world real and pretend flow together.

In my world I'm just learning to keep them apart.

Later

I heard the phone ring again and Mrs. Duncan came to my door. She had a funny look on her face when she said the call was for me, and I wondered why. The man who called said his name was Mr. Borton, attorney for the young couple who want to adopt my baby.

"I hope I haven't embarrassed you," he said. "I forgot you are living under an assumed name. I'm afraid I used your real name." I could feel my cheeks burning, but I told him it didn't really matter. He asked how soon I could come to his office, and I said Tuesday, when I go to town for my checkup. He said, "See you then," and I hung up and turned to face Mrs. Duncan.

"Isn't Jean Thompson your real name?" It sounded like an accusation. I just stood there staring at her, like a fool. All I could think of was that Miss O'Connor had told me the wage-home mother wouldn't ask questions.

"Don't you trust us, Jean? We want to be your friends." I could only nod and stammer, "I—I know you do—of course I trust you," but I couldn't make myself tell her the truth, and I just stood there till she finally smiled and said, "We know it must be hard for you."

I smiled back as well as I could and turned to go to my room. I hope she never gets back on that subject. Maybe I can trust her —maybe I can't. But she would be one more person who knows the truth, and she would tell her husband. I can't take chances.

Near midnight

Maybe this was the wrong time to do it, but I wrote Patrick anyway. I have to have someone to talk to—even if it is only on paper. I didn't tell him where I am and why, but I did write about how I feel. About being alone and afraid. About how hard it is when there are no absolutes, no easy answers. How hard to grow up to the big world, the excitement, the new horizons, ideas, people, unlimited hopes, unlimited dreams—unlimited reality.

And then you can't go back to the security of being very young, assuming that grownups know all there is to know, even if you don't. And you can't go back to wipe out your mistakes and what you've learned and done. You're so damned alone—you can't share your innermost thoughts, even if you know them. You can never quite reach the soul of another person, and you're even afraid of coming close—afraid you might get hurt, lose something of yourself.

It is the sort of letter I never send if I wait till the next day to read it again. But I won't read this one again. I want to share something with someone.

Tuesday evening, the 13th

What a day this has been. So much has happened in a few short hours, and I had grown so used to the peaceful solitude of this valley. Today shattered that peace and I'm really not sorry.

To begin with, Mrs. Duncan drove me to the Home this morning. We didn't say much on the way; there is tension between us and I don't know how to ease it without telling her about myself.

She stopped the car outside the Home and handed me a check for the money I have earned. My cheeks got red.

"I'm sorry," I said. "I haven't got any identification, I can't cash a check."

"Not even in there?" She looked surprised and nodded toward the Home. I shook my head, and without a word she gave me twenty dollars from her purse.

"My husband will pick you up at six," she said.

"Thank you, I'll be ready." I got out of the car and waved good-bye. She didn't look back.

I could hear laughter from the second floor of the Home. The windows were open and the old house sounded like a happy place. I walked slowly up the steps and through the green door. A new girl was waiting on the couch in the hall, her hand on the handle of a worn suitcase. She looked up at me quickly with that guarded look we all carry in here from the outside.

"Poor kid," I thought, and smiled at her. She smiled back. It felt good to be a visitor, to nod to the girls I saw passing in the hall and watch them grin when they recognized me as one of them. Then Mrs. Warren stuck her sour face out the window from the inner office.

"Go straight to the hospital floor, Jean," she said sternly. "No outsiders in the dorms, you know." I felt cheated and left out. "The hell with her," I told myself, and marched right upstairs and into the green dorm, where Karen was knitting on her bed.

"Hi, there!" I yelled, so loud she jumped. Her face lit up when she saw me.

"Hi, stranger, join the party." Karen waved a chocolate bar in my direction and then she noticed that I was wearing my blouse loose over my skirt instead of tucked in and with a belt.

"Why Jean," she said with mock seriousness, "if I didn't know you better I'd suspect you were pregnant." We laughed and

Karen patted her own big belly. "Guess what, the doctor says I'm engaged, I'm going to be two soon."

"Wow!" I said, playing along. "Who's the lucky guy, anyone I know?"

"Stupid!" Karen laughed. We all use the old joke about being engaged. Actually it is the term the doctor uses to say that the baby's head has come down into the upper part of the birth channel during the last stage of the pregnancy. I sat on the edge of Karen's bed and munched on the chocolate bar. The noises of the dorm—two radios playing at once, voices talking and laughing—seemed pleasantly familiar.

"You missed seeing Marian," Karen said. "She left yesterday and she'll be married Saturday. She said to tell you good-bye. You never saw a happier bride-to-be."

We sat quietly with our own thoughts for a minute. I thought how hard it must have been for Marian to make her decision to go home. How she had to come all the way out here to a home for unwed mothers to get away from the pull of those who love her and want to make up her mind for her.

"Don't forget our date," Karen said. "Coffee at Toni's."

The intercom made noise: "Jean T., Jean T., you're wanted on the hospital floor." I ran down the hall where Nurse Simpson was holding the white hospital gown for me.

"Doctor can't wait, hurry up!" She didn't sound mad, so I guess she knows how it is to come back and meet friends after being alone in a wage home.

The doctor listened to my heart and then put the stethoscope against my tummy, listening intently.

"He's there, all right," he said. "A good, strong heartbeat." I felt quite excited.

"When will it be?" The doctor smiled a little at my eagerness. "It's a bit early to tell yet. When are you due?"

"In March."

"Maybe you'll be just a little bit early." He smiled.

"Thanks a lot." I jumped off the table and headed for the door.

"You're welcome," he said after me, "but don't blame me if you're late."

Shucks, I said to myself, I won't be late.

Karen was waiting for me, all ready to go, but I had something to do first.

"Why don't you go ahead?" I said, ignoring the question mark written all over her face. "Tell me how to get to Toni's and I'll meet you there."

"Can't miss it," she said. "It's on your left when you walk down Main Street. There are always a couple of big trucks parked in front."

"O.K." I waved her out the door and walked down the hall to Miss O'Connor's office. She looked up and smiled when I opened the door.

"Come right in, Jean," she said, pushing away a stack of papers on her desk. "What's on your mind?" I dumped down in her red chair and pulled on the cigarette she offered me as if there were some kind of hidden courage in the smoke.

"I'm on my way to see a lawyer and I need some advice," I said. Miss O'Connor watched me calmly.

"Oh?" she said.

"I've decided to have the baby adopted, and I want an independent adoption," I said, watching her face for a reaction. But she just kept on looking at me with that quiet smile of hers.

"So?" she said. "What do you want me to tell you?"

"The people who want my baby—I'm seeing their lawyer for the first time today—they've already adopted one baby three years ago. I thought—I mean, they've already been approved by the state." I broke off, feeling angry at my own clumsy speech. "Why don't you tell me an agency adoption is better and safer?" I said, hoping she'd say something, try to change my mind or something. "Why don't you tell me I run all sorts of risks with a private adoption? Oh, and I haven't told you the clincher yet—I want to meet them to make sure. . . ." Miss O'Connor smoked in silence, watching me punch out my cigarette in her ash tray.

"I'm not here to tell you what is better for you," she said quietly. "I'm not here to make up your mind for you, even if you'd just love to have someone tell you what to do. However, I can tell you what is involved in what you want to do, once you have made up your mind to do it. Why do you want an independent adoption, by the way?"

"I want to make sure the baby has a mother to go home to right from the start. I just know a baby needs a mother right away," I ended lamely.

"Of course, you are aware of the risks," Miss O'Connor said briskly. "You, the natural mother, are actually legally responsible for that baby until the court closes the case six months after placement. The adoptive parents could change their minds, you know, in case something unforeseen develops."

I nodded. "That's why I wanted to meet the couple," I said, "to be sure they were the kind who would go ahead regardless." I moved uneasily on the chair. "Is it common for natural mothers to meet the adoptive parents?"

Miss O'Connor took a book from her shelf. "This is the California law governing such cases," she said. "Actually, in independent or direct placement adoptions, the natural mother has a right to know the names of the adoptive parents and to assure herself, either through a third party or by personal contact, that the adoptive parents satisfy her requirements."

"But agency adoptions are safer, aren't they?" I don't know what made me ask. Miss O'Connor smiled.

"That depends on what you're looking for," she said. "The agency *does* take over full legal responsibility for the child until placement is made, and in an agency adoption the family is thoroughly evaluated by the welfare agency before placement of the child . . . just to give you an idea," she said. "A few years ago, of nearly ten thousand adoptions in California alone, nearly seventy per cent were independent, and fifty per cent of those were open agreements—that is, the natural mother met the adoptive parents. Risk or no risk"—she smiled—"you aren't the first one to take it." I got up to leave.

"I came to get some moral support, if nothing else," I said. "I'm scared to death of making this thing final!"

"I bet you are," Miss O'Connor said. "You wouldn't be normal if you didn't have some serious thoughts about doing this. But I will say one thing before you go. I am confident that you will do the right thing for all parties involved. Come back and see me soon." She reached her hand across the top of the desk and shook mine.

"Good luck, Jean," she said, and in spite of myself I had to smile back. "Thanks," I said. "See you later."

I found Karen at Toni's. She lifted her coffee cup in greeting when she saw me and said, "My fifth. What kept you?"

"Nothing much." I tried to sound casual. For some reason I couldn't come right out and tell her about my talk with Miss O'Connor. Karen shrugged her shoulders, but her smile was the old friendly one. "You keep your secrets, I keep mine," she said. "Nine months of secrets and for me it's almost over." She nodded toward the Coke-calendar on the wall.

"See that?" she said. "I've been sitting here staring at it. I can hardly believe my due-date is this close. You know, sometimes I wake up in the middle of the night thinking I'm going to stay pregnant forever. The baby won't come out or I'll be discovered or something."

"I know," I said, wondering if March would ever come. Toni

brought another cup and the coffee pot. He is short and bald.

"This is Jean," Karen said. "She's a new one on the Hill and she'll be drinking your coffee after I've gone, won't you?" I nodded and smiled, and Toni grinned all over his round face. "You girls are always welcome here," he said. "I've got plenty of coffee." When Toni had left us Karen stirred her coffee absent-mindedly.

"You know, now that it is almost over, I think the worst part of it has been the damn loneliness," she said. "I've adjusted pretty well to the hiding and the lying to the outside world, but I've just never gotten used to being all alone inside." She shook her head, then straightened up and smiled. "But a lot of good has come out of this, too, you know," she said. "I've discovered who my friends are. You find out when you're in trouble. Some people you thought would stick with you always simply walked away, and others, the ones you'd never even thought would care, want to help."

"I used to wonder how you could laugh and joke all the time," I said. "The day I met you and Lucy, I was almost shocked."

Karen laughed. "All the new ones go through that stage," she said. "I did, too. But after a while you find that crying isn't going to help any. You know you've got all that time to wait before you can go out and do something else, and you might as well laugh."

"I know," I said. "But it still upsets me. I think maybe if you learn to live with something that hurts and almost ignore it, then you're not going to try to make it better."

Karen looked at me and shook her head sadly. "You can't do much about it now, Jean. And don't try to be a crusader. Wait till you're out of here, anyway."

"It's just that I keep wondering why I did it in the first place," I said, "and just putting the whole thing out of my mind isn't going to answer the question."

"Forget it, Jean." Karen sounded almost angry, then she smiled and put her hand on my arm. "Look, kid," she said, sounding like a big sister. "This isn't the time to do all that analyzing of yourself and everybody else. It will only drive you nuts and you won't be any wiser. Believe me, I've been the route." She sighed. "But maybe you've got to find out for yourself—is that it?"

"I don't know." I shrugged my shoulders.

"Well." Karen sounded resigned. "I've thought about it a lot, too, and I've only figured out one thing for certain. That the bunch of us in here got pregnant because we had some other big problem we just didn't know how to handle. All of us had a different problem, but one that had us in a corner and pretty miserable." She lit another cigarette. "Of course, I may be all wrong." We smoked in silence. I wondered how many lonely hours and how many cups of coffee and cigarettes it had taken Karen to get to that point. Then I noticed the time—it was nearly one o'clock and I would be late if I didn't hurry.

"I've got an appointment with a lawyer downtown," I said. "Gotta run."

"Lawyer?" Karen looked curious.

"Independent adoption," I said, and the look on her face made me glad I'd said it. Karen held out her hand and I took it.

"Good luck to you," she said. "Give me a call soon."

I watched her go up the sidewalk, her head high and her shoulders straight under the trench coat. She walked with an easy stride in spite of her heavy stomach; she looked very sure of herself. I turned and walked slowly the other way, toward downtown. There was a faint smell of salt from the ocean and I was suddenly aware of how much I had missed the sea breeze in the hills. I've never liked big cities. I get a feeling of being trapped among those big buildings and all the people rushing around for no apparent reason. But this afternoon I didn't mind it. I was an outsider, and what I saw and heard and smelled had nothing to do with me. I thought it was like listening to a great symphony orchestra. The listener can turn and walk away at any time. But if he is one of the musicians he can't, and his little sound is indistinguishable from all the other sounds. (Unless, of course, he plays off key. Then he gets all the attention his little ego might want.)

I found the lawyer's name on the directory in the lobby of one of the big downtown buildings. I couldn't think of another excuse to keep from facing the interview. Seventh floor. In the elevator I stared at all kinds of necks in front of me—thick and thin ones, close-shaven ones and long-haired ones. I've always looked at people's necks. We are all so clever at making masks you can't tell much from faces anymore. Even eyes lie shamelessly. But hands and necks say a lot.

I suddenly had a ridiculous thought. There we were, bodies in a small cubicle, our minds somewhere else. I thought of invisible lines going from each of us to parts of the city and all over the world. What if the elevator fell down and killed us all? Would our thoughts still be floating around somewhere, without purpose, and nowhere to come back to?

"Seven," called the elevator operator. I pushed past the necks and followed a plump lady with tinted red hair down the corridor. She walked straight past the door with the legend, RICHARD E. BORTON, ATTORNEY, and I had to go in alone. The carpet was thick and green, and the young well-groomed lady behind the mahogany desk looked at my shabby trench coat and my only pair of low-heeled walking shoes splashed with muddy rain.

"I think Mr. Borton is expecting me," I said, fighting the impulse to turn and run out the door again.

"Oh, you must be Miss Thompson," she said, and flipped a switch on the little box on her desk.

"Miss Thompson is here to see you," she said in a warm

smooth voice. The door to the inner office was opened quickly and Mr. Borton came toward me with outstretched hand and a tremendous smile. He was wearing a tweed suit and red tie, and looked about forty. There were discreet streaks of silver in his thick black hair. His voice was booming.

"Hel-lo, Miss Thompson. I'm so glad you could make it. Come on in." I smiled weakly and he grabbed my hand, almost pulling me into his office. I sank into a huge, soft, leather chair and Mr. Borton sat down behind his gleaming desk in his red leather swivel chair. I waited for him to start the conversation. He cleared his throat several times, tapping his finger tips lightly together.

"Let me make it clear from the start," he said, a bit awkwardly, and I suddenly felt like laughing. "I don't hold your present situation against you . . . it can happen to the best of us." I smiled and thought his clients had better not be anything like their lawyer or I'd call the whole thing off.

"Well, then," Mr. Borton continued after a slight pause, "my clients have informed me that you are interested in making arrangements with them for the placement of your child." I nodded.

"I want everything settled as soon as possible," I said, "so that when the baby is born there will be nothing further to discuss."

"I can understand that," he said, "and I'm sure the adoptive parents will agree." He reached for a pad and a silver pencil on his desk.

"Some questions. The adoptive parents would like to know your age, occupation, the father's age, race and occupation." I rattled off the answers; I'm getting used to them by now.

"I'd like to meet them," I said, and Mr. Borton stopped taking notes to stare at me.

"I—I really don't know," he said. "I've only met them over the phone—they may not agree to a meeting." I drew in my breath sharply and felt the nervous quivering in my stomach.

"It's very important to me." I tried to keep my voice calm. "And it is quite common here in California." I nodded toward his bookshelf where I could see the book on the California code Miss O'Connor had quoted. "It's in your book there"—I spoke quickly—"that in independent adoptions the decision is made after the two parties—the natural mother and the adoptive parents—have obtained a thorough knowledge of each other and have come to a complete agreement." I could hear how stilted my own words sounded, but Mr. Borton was listening intently.

"As the natural mother I have a right to know the name of the adoptive parents," I continued. The nervousness had left my stomach entirely. "I saw some statistics in my social worker's office. A few years ago fifty per cent of the independent adoptions in this state were open agreements—the natural mother met the adoptive parents."

Mr. Borton was smiling a little. "You've done your homework, haven't you?"

"It's my baby and my life," I said. He leaned forward across the desk, his voice businesslike again. "I'll call the young couple today," he said. "I'll tell them what you have told me—I hope they will see it the same way."

"If they don't, I'm afraid I can't go ahead with it," I said. "I've got to be sure in my own mind that they are the right family for the baby."

"I'll do my best to convince them." Mr. Borton got up from his chair and held out his hand. I got to my feet. We shook hands and he said he will call me as soon as he has talked to the adoptive parents. I caught the bus back to the Home and got there just as Mr. Duncan drove up to pick me up.

It's way past midnight and I've written half a book, it seems. But I remember Mikki's words—write it down, then you can go back and read how you felt. I'm terribly forgetful. Is it a blessed forgetfulness? Or a dangerous one, just putting off facing things?

Wednesday the 14th

Another letter from Patrick came today. He writes as if he can read my thoughts: "Yes, you might get hurt if you share your innermost thoughts with someone. But without sharing you will never live fully—the way I believe we are meant to live. Our masks of lies and pretenses become so much part of us that two who could love each other may meet and pass by and never know this."

He is right, of course. But I don't dare share with him now. I have to write about little things, walking in the rain, a favorite book. I can't say where I am and what I'm doing. I am not lying to him. I just avoid the truth.

Friday the 16th

Late at night when the house is quiet, I talk to myself. Is it any different from writing to myself in these pages? Or I talk to my books or to the pictures of Mother and Dad. It's good to get thoughts off my chest, hear them spoken out loud, even if they can't hear me.

I remember last summer. I helped Mother clean the attic and we found the baby clothes she has saved all these years. Mother held the little dresses and shawls up for me to see and said I could have them for my first baby. I wonder how I'll feel when I put them on my second-born—if I ever have another child.

I've promised myself I will try not to think of things that may make me break down. I break that promise often.

Sunday the 18th

Last night I kept waking up to go to the bathroom. When I returned to my room after the last trip I could see the faint glow over the hills where the morning was about to happen, and suddenly I felt a light fluttering inside my stomach, a tiny movement, almost like a gas pain, only I knew it wasn't. I stood still and it came again.

My baby moved.

Friday the 30th

I'm halfway through the pregnancy, the calendar says. Four and a half months to go—that isn't all eternity, even if it seems that long since I first thought I was pregnant. I alternate between wishing it were all over and wishing it would go on and on and on. When I try to imagine what it will be like to give up a baby, it is still only in my mind, and that baby is safely inside me. Four and a half months from now it won't be imagination anymore. I believe that somewhere inside of me is the strength I will need to face that test, but I'm scared to death that I won't have the self-discipline to call on that strength. If I chicken out then, all this will have been for nothing.

I ask myself the same questions over and over again and repeat my own made-up answers. I must get off that merry-go-round and think of something concrete instead—sew the tear in my shirt, make a pot of coffee.

NOVEMBER

Saturday the 21st

Less than four months to go, and it's suddenly crowded inside my skin. My ribs are sore, and when the baby kicks I think maybe he or she is trying to tell me that two of us are just one too many inside a skin made for only one. I feel stuffed every time I eat, and I can't drink anything at night without having to go to the bathroom every couple of hours. If I could only walk in my sleep. Now I toss and turn and chase my thoughts in hopeless circles most of the night.

I don't want to be this tired. I can't think straight when I don't get enough sleep. I feel sometimes as if someone or something is trying to see how far they can push me before I break. But I'll get through this thing one way or the other!

And that's a rotten lie. I'll never be through with this; it will be part of me always. A child growing up somewhere—and my name on the adoption papers.

Later

The phone rang—for me. Mr. Borton called to say the adoptive parents have agreed to meet me. Their name is Carver, Martha and Don.

"They would like to have you come to their house tomorrow afternoon," he said.

"So soon?" I could barely whisper.

"You asked for it, didn't you? I had to work hard to convince them." Mr. Borton sounded impatient.

"Thank you. I appreciate it," I said, and felt as if a web were slowly closing around me. But that's what I wanted, to have it so tied up I can't chicken out.

"Mr. Carver will pick you up after church tomorrow," Mr. Borton said. "And by the way, I suggested he call you by your first name, and that you pretend you know each other—we don't want to embarrass you any further with the Duncans." The poor guy is probably sorry he made that silly blunder the first time he called.

Saturday night

I'm scared. What if I don't like them? Or, if I like them a lot, is it going to hurt more knowing what they are like and what their house is like?

Karen just phoned from the mothers' ward. "I made it," she said. She was bursting with excitement. "We get to make one phone call right after we get back from the delivery room. I just had to tell someone right away—did I wake you up?"

"Of course not," I said. "It's only ten thirty. I'm a night owl, you know—and congratulations!"

"Jean, I'm so happy!" Karen's voice trembled slightly. "You ought to see me—no tummy and I'll start exercising tomorrow morning. I'll be out of here in a week."

"That's great," I said. "I'll be coming in Tuesday for my

checkup—I'll expect you to be doing the twist by then." It was a silly thing to say but Karen laughed.

"I'll show you. And I'm gonna get a solid night's rest, on my tummy, starting right now. Goodnight, Jean."

"Goodnight, Karen."

So she's over the line now. I wonder if she had a boy or a girl. I didn't want to ask and I guess she didn't want to tell me.

Sunday morning, the 22nd

It's a lovely morning. The sun is gold over yellow and brown hills. I am all dressed up, which isn't so much—just the black skirt and my red-and-green top; it isn't as washed-out as the others. I've shined my one pair of shoes. I don't want to make an altogether bad impression on Mr. and Mrs. Carver. He ought to be here any minute now. Oh God, I know I wanted this. Give me strength to go through with it.

Sunday night

I am really too tired to write about it now, but I'm afraid of going to bed. I'm afraid of turning out the light and trying to go to sleep, and starting to think. So maybe I better talk about it, get it all down on paper and out of my mind.

Don Carver came to get me a little after noon. He is tall, blond and good-looking. I don't know what I had expected but I hadn't thought he would be so nice. He took my hand in a firm shake and his blue eyes looked at me calmly. Mrs. Duncan was watching us, so I smiled and said, "Hi, Don, nice to see you."

The Carvers live in a suburb on the other side of town and the ride took nearly an hour. I felt awkward at first, but Don

Carver must have felt worse. He was trying to make polite conversation.

"How do you like California?" he asked me, and then bit his lip. I guess he thought it wasn't the thing to ask under the circumstances.

"It's nice," I said. "I'd like to come back on a visit some time." I caught his quick glance and realized I'd said something wrong, too. We rode in silence for a while and I was beginning to think the whole thing was a terrible mistake.

"I'm sorry," I said. "I—I just had to see you."

"We understand," he said. "It took me a while, but my wife said she thought she understood how you must feel—and we do want another child very much."

"When this is over," I said, "try to believe me when I say I will honor the decision we make together. I won't try to get in touch with you again. I don't want you to get in touch with me. I only want what is best for the baby." I could see Don's knuckles whiten over the steering wheel.

"I believe you," he said, "because I want to believe you. We know this is harder for you than for us. We have nothing to lose —and a lot to gain."

Don turned into a quiet side street with trees lining the sidewalks and stopped in front of a brick house. The white front door flew open and a little girl ran down the path to meet us.

"Daddy, Daddy!" She screamed with delight when Don lifted her high in his arms.

"Hi, princess," he said. "Say hello to Jean. This is our Kathy."

"Hi, Jean," Kathy said politely and turned immediately to Don again, her red curls dancing around her bright little face. "We're gonna have cookies and tea—Mommy's got it ready." Don laughed and put her down on the sidewalk. She held his big hand and they walked together to the house. Martha Carver was standing in the doorway. I was almost afraid to look at her, but her eyes were warm and when she smiled there was a dimple in her cheek.

"I'm glad to meet you." She took my hand and I could only smile and whisper, "Thank you." I didn't trust my voice.

The tea table was set for four in front of a blazing fire in the fireplace. Kathy's little rocking chair was there and she ran to sit down before we did. I glanced around the room—the books, the paintings lining the walls, the worn but comfortable furniture, the bay window full of potted flowers. Martha caught my glance.

"It's one of our hobbies," she said, "growing flowers."

"How about some tea?" Don pointed to a comfortable chair. "Sit down, Jean, make yourself at home." I did want to feel at home, and it would have been so easy in that friendly room with those nice people—if only I hadn't been the mother of their future child. Half a birthday cake stood on the tea table.

"Kathy's birthday was yesterday. We thought we would celebrate two days," Martha said.

"How nice," I said. "How old are you now, Kathy?"

"I was three yesterday," said Kathy, "and I'll be four soon."

Martha poured the tea and Kathy drank hers with half a cup of milk and sugar. Her dolly sat on her lap and had to be scolded for spilling crumbs on the rug. Over Kathy's head we talked about the weather, getting ready for Christmas and the latest news, carefully avoiding any mention of babies or related subjects. At last Martha turned to Kathy with a smile.

"Time to put baby-doll down for a nap, Kathy." Kathy looked reluctant but her mother's face was stern. She stood up with her doll in her arms.

"Baby-doll isn't very sleepy," she said. "But I'll read her a long story so she won't cry." We sat in silence till Kathy had left the room.

"We know your decision is a hard one," Martha said, looking directly at me. "We want you to be sure in your mind about us—

so ask any questions you like." I searched for words—what could I ask them? How could I even try to compare what I have to offer a baby with all they have to give?

"Now that I'm here, the questions don't seem so important," I said. "I wasn't concerned about your social or financial situation. I know you've already adopted one child and have gone through the state investigation. I just wanted to know what kind of people you are." I hesitated but Martha's eyes urged me on. "I really have no right to question you," I said. "I don't have anything to give the baby myself. But there are some things I wish I could give—maybe I'm selfish in wanting you to give them."

"Not selfish," said Martha. "You have a right to look for what you think is best for your child."

"I'm also asking for my own sake," I said, "so that if I have regrets in the future they will be for myself, not because I worry about the baby." I was beginning to feel terribly self-conscious. "Maybe what I really want to know is whether you're going to be good parents and love the baby. And how can you ask a question like that?"

Don smiled. "I can tell you that I think Martha is the best mother in the world. Or maybe you should ask Kathy. They sing together and read and grow flowers."

"No one is a perfect parent," said Martha. "We can't promise happiness for your baby. But we would love him—or her, perhaps even more than if we had had our own children, because then we would not have known how badly we wanted them. We would have taken them for granted." Tears kept flooding my eyes. I had been scared of coming, scared of finding out for sure, but now I knew.

"You're more than I had hoped for," I said, trying to keep my voice even. "My baby will have a wonderful home here with you." I could see the tears in Martha's eyes, too.

"Do you mean that?" she said. I nodded.

"I'm quite sure. And I'm happy, believe me. You need a baby and my baby needs a home. It is really the best for us all." I thought if I didn't get away from there I would break down.

"Tell Mr. Borton to draw up the papers and have them ready to sign as soon as the baby is born. I'll call you when it gets here. It should be around March seventeenth." I thought I sounded quite businesslike.

Martha dried her eyes. "I'll tell my pediatrician," she said. Don offered to drive me home but I thought I would like to be alone, maybe go to the movies and catch a bus back to the hills later.

We said good-bye at the door and I walked quickly down the sidewalk without turning to look back.

I walked the two miles to the business section, and when I got there I didn't want to see a movie after all. The two dollars in my pocket was all the money I owned. I can't get used to being broke all the time. I was broke before this all happened, but it never bothered me then. I knew I could always get a job. Now I'm trapped. I'm not free to earn my own keep. The ten dollars I get at the Duncans' feels like charity, not wages.

My feet began to ache, and inside I felt numb. I could see my reflection in a shop window—the trench coat doesn't quite hide the bulge of my stomach anymore. I wanted to keep on walking, keep from thinking about the baby and Martha and Don Carver and the things I had said to them. The nice noble things about wanting the best for the baby and how I thought they could give him that. It's true, but I don't want it to be.

I sat down on the bench by the bus stop and closed my eyes against the glare from the sinking sun. And then I heard a voice —a trembling old voice. I looked up and saw an old lady sitting next to me, her yellow cheeks painted bright red with rouge, her white hair in thin curls under a pale blue hat. As pale blue as her watery old eyes, so sad and lonely, staring straight ahead while she talked to no one.

"I'm so tired. Where I come from the altitude is so much higher—" Her voice trailed into a whisper and she turned her head slowly to look at me. I saw her hands, limp inside the clean white gloves in her lap. I couldn't make myself speak to her. From inside my loneliness I saw hers—and I got up and walked away.

I walked into the drugstore on the corner, where the juke box was blaring and some boys in leather jackets and high boots played the pinball machine while their girls—also in leather jackets and boots—were dancing with each other. I bought some chewing gum and read the titles of the paperback books on the rack. The monotonous beat of the music drove me out of there, and I went into a small Chinese restaurant right next door. On the wall was a bright poster advertising Coke as "The Friendliest Drink in the World." It made me feel very lonely and I wanted to cry. I ordered tea and the little Chinese woman brought fortune cookies with it. I broke the cookies and two of the slips of paper inside them said that I'm lucky in money matters and will soon hear pleasant news from an unexpected source.

I walked back to the bus stop and the little old woman had gone.

Tuesday night, the 24th

I went in for my checkup today.

At the Home they are getting ready for Thanksgiving. They make the most of every holiday to break the monotony for the girls. Gay streamers decorated the hall and on the chapel door someone had tacked up a copy of the first presidential Thanksgiving proclamation of 1789.

> . . . It is the duty of nations to acknowledge the providence of Almighty God, to obey his will, to be grateful for his benefits, and humbly to implore his protection and favor. . . .
>
> *George Washington*

I met Mrs. Warren in the hall and she wagged a finger at me. "No visiting on the dorm this time, Jean." I smiled and shook my head. There wasn't anyone I wanted to see in the dorm. Karen is in the mothers' ward.

Doctor Norwad was on duty again. I hope he'll be on call when I get ready to deliver. All the doctors who take turns serving our hospital are nice, but most of them make us feel like numbers on a production line when we jump up on the examination table, one after the other. Doctor Norwad is different. He makes you feel important. He cares. He even has time to answer our questions and explain how we are doing in words we can understand, instead of the mumbo jumbo the nurse puts on our charts.

He felt my tummy. "Long and narrow," he said. "Floating, dipping lower right quadrant," which means in plain English that the baby is in correct position, head down but not engaged yet. Doctor Norwad told me all about it, smiling, his blue eyes circled with dark from lack of sleep.

I'm holding my weight down, too. Nurse Simpson is pleased. Heaven knows I fight the urge to nibble all day long. Smoking isn't a healthy substitute but I can't stay off cigarettes, too.

I saw Karen in the mothers' ward. She's up already and looked slim in a housedress. But her smile wasn't as bright as it used to be and she joked about having the "after baby blues"—literally.

She told me she signed the papers yesterday and the baby will go to a foster home tomorrow to await placement in a permanent home.

"Thank God she's healthy," Karen said. "Healthy white babies find homes quickly." We joked about my big tummy, and I told her I met the adoptive parents Sunday.

"Good for you," Karen said, and we were silent for an awkward minute.

"Well, I'll be going back to the snow country the day after Thanksgiving," Karen said brightly. "It'll be good to try on skis again."

"Think of me when you go down that slope," I said, and Karen laughed. "How can I ever forget!"

We sat in silence again—it was hard to think of something to say. There was a wide gulf between us and nothing to bridge it with. I think we could have been good friends on the outside, too, but Karen never told me her real name or where she comes from, and I never told her mine. If we meet again it will be by accident—and we may not even want to admit we ever knew each other. I got up to leave.

"So soon?" said Karen, but in her eyes I saw the relief I felt, too.

"Mrs. Duncan is picking me up early," I said. I held out my hand and Karen took it.

"Good luck, Karen," I said. "It's been good knowing you." My throat felt funny.

"Good luck to you, Jean," Karen said, and the look in her face brought the tears to my eyes. I pulled her close and hugged her quickly, and then I ran out the door.

I think I understand now why we joke so much in the Home and don't get too close to each other. It hurts to be alone—but it's even worse to part from someone who could have been a friend.

Thanksgiving evening

The house has been full of the smells and sounds of Thanksgiving all day. Mrs. Duncan told me I could have the day off, but I have nowhere to go, so I went for a long walk among the hills and I've spent the rest of the day in my room reading. I borrowed Hemingway's *The Old Man and the Sea* from the shelf in the den. I remember reading it several years ago and thinking that it was very good, but now the story of the old man—alone with his pride and his pain and his giant fish, battling the sharks who nibbled his catch away—means so much more to me. Mrs. Duncan just knocked on my door and asked me to join them for Thanksgiving dinner.

The table was beautiful. It is tradition in the Duncan family that all members of the household help prepare the meal. Peter and Lori had helped cut up vegetables and decorate the pumpkin pies. Mr. Duncan had made the dressing and stuffed the turkey. Now they were dressed in their Sunday best, and the candles were lit. Mr. Duncan asked us all to join him in prayer. We bowed our heads and they all prayed together. It must be their traditional Thanksgiving prayer.

> Bless our house, Lord, and all in it.
> Bless the food you have given us.
> Bless those who suffer and those who are in need.
> For a wonderful year—we thank you. Amen.

"And bless Grandma and Grandpa and Jean," said Lori, peeking at me. I could feel the silly tears coming and I didn't want to make a fool of myself. I pushed back the chair and got to my feet. They all stared at me and I tried to keep my voice steady.

"I don't think I can eat—will you excuse me please?" I left

the room before anyone could say anything, and I felt like a heel. Why can't I accept kindness? Why do I turn away when someone wants to help? I have so much to give thanks for (or is that proud ego of mine too stubborn to say that sincerely?).

Thanksgiving last year—Mother and I were alone. My brother had joined the Air Force and Dad was caught in a snowstorm in the Midwest where he had been giving guest lectures at a college. What a miserable Thanksgiving it turned out to be.

I had come home from college and had hoped that the months I'd been living away from home had made things easier, but in the big house, that seemed so empty without the rest of the family, Mother and I ate our Thanksgiving chicken dinner in polite silence. I found that we still had very little to say to each other even if Mother always talked a blue streak. She bored me with her fussing over my looks—"You're too thin—do you eat regularly? Shouldn't you wear a heavier coat?" Maybe she meant well but I couldn't take it. Funny, now that I look back on the whole thing, I guess I missed Daddy. We sure do a lot of arguing when we're together and we clash on practically every subject in the book. But it's better than Mother's endless small talk.

Maybe I'm crazy, but right here, right now in my little room, I feel more like giving thanks than I did last year.

Saturday the 28th

The hills were different this morning. It is warmer. Clouds were drifting across the sky and their shadows moved over yellow dry hills. I stopped to wonder what the difference could be. The wind, maybe? It seemed to come from a different direction. There was waiting in the air—something like April in the north when spring comes. Except here it is November in the southwest.

When I went to get the mail the rain came. The drops fell with a rustle against the dry leaves and the heavy scent of dry grass rose from the earth like a soft wave. The drops fell on my face and my bare arms and I felt that, like the hills, I had been waiting for the rain.

I went back to my little room just to sit by the window and watch the rain and the mist drift through our valley. I was never able to just sit before. I would feel guilty. I would feel I ought to be doing something or going someplace.

Here I can sit and it is wonderful. I know this peace doesn't fill all of me, but I am grateful when I have it. It takes little to throw me off balance, but each week, here in the valley, I feel I am getting stronger.

DECEMBER

Tuesday the 1st

It is the first day of the Christmas month and already the holiday spirit is beginning to haunt me. I went in for my checkup today and the group social worker, Mrs. Horner, caught me in the hall to ask if I want to help on the Christmas Service project. This year the girls are going to repair and dress dolls for the Salvation Army Christmas Welfare.

"And of course you're welcome to join our Christmas workshop every afternoon this week and next," she said, holding on to my arm. "We have a wonderful selection of material to make gifts for your friends and relatives—jewelry, wallets, aprons, stuffed toys and lots more."

"I can't get transportation," I said. "I'm too far away to come by bus."

"Oh." She looked disappointed. "Well, you'll enjoy working in the crafts room when you come back to stay, I'm sure," she said, more to herself than to me.

Actually I *would* like to make some gifts for my parents. I used to make gifts for them when I was a little girl. I remember a pair of socks I knitted for my Dad—they were too wide around the ankles and too long between the heel and the toe—but Dad wore them every night during Christmas week. This year I would like to give them something special, and I don't even have the money yet to shop in a five-and-ten.

Friday the 4th

I walked the two miles to the crossroads where there is a small drugstore, a grocery store and a gas station. For my ten dollars I bought a scarf for Mother, a lighter for Dad and some Christmas cards. I've wrapped the gifts and written the few greetings I have to send. I hadn't thought it would hurt so much to write those greetings. Happy, easy words, and I wanted to say so many things I can't say.

I have this feeling of closeness to Mother and Dad that I can't remember ever having felt before. Sure, we haven't gotten along very well the last few years, but what was it Martha Carver said? "No one is a perfect parent—but we'll love our children," or something to that effect. I guess Mother and Dad in their own hardheaded way love me a great deal, and maybe I haven't been the easiest daughter to rear.

I think I'm getting overly sentimental. Those Christmas carols on the radio really get to me.

Sunday the 6th

Dorothy was here today! She didn't phone, just drove up to the house and rang the doorbell. I happened to answer it and my heart skipped a beat when I saw who it was.

"Hi, there," she said, her blue eyes laughing. "Aren't you going to ask me in?" Then she held out her arms and when she hugged me I wanted to burst with happiness. I didn't know how much I had missed her. I showed her to my room and she stood in the center of the floor, turning slowly, seeing the sunlit walls, the flowers in the vase on my table, my miniature Christmas tree hung with the balls I made from tinsel and yarn.

"This is nice," she said, and I saw tears in her eyes. She smiled. "You look wonderful, Jean." I remembered the day she came to my dark hole of an apartment.

"You don't know how I've worried," she said. "I've imagined you pale and panicky."

I said, "Let's have some tea and cookies," and she waited while I brought the tray from the kitchen. I baked the cookies yesterday, leaf-thin spicy ones, from an old recipe my mother taught me years ago.

"I must have known you were coming," I said. We talked for nearly four hours. I told her about the fumbling lawyer and about meeting Don and Martha Carver. It was wonderful to talk to someone without pretending, without having to watch my words.

"I'm glad you made the decision this early," said Dorothy. "I'm quite proud of you, you know." It made me feel good to hear her say it but I had to tell her that sometimes I feel very bad about it. Sometimes I still want to keep the baby even if I know that is all wrong.

"You've come a long way," Dorothy said. "You will find the strength. Don't worry." Dorothy's voice and laughter drove away the shadows of the months of loneliness and fear and lies. I had almost forgotten that it is normal to relax and to laugh and to say what I really mean. To be whole again instead of a split personality. I kept wishing I could stop the clock and hold on to these hours. With Dorothy here, my little room was suddenly no longer isolated from the rest of the world.

It was hard to say good-bye, to hug her and know that the loneliness will close in on me worse than before. Dorothy will be back in March. She has promised to carry the baby from the Home to Martha Carver. Then we'll travel back east together.

Meat loaf for supper. Peter and Lori emptied their plates and I forced myself to eat. After dinner I did the dishes slowly, stretching the time. I made dough for tomorrow's cookie baking. The girl who was here from the Home before me, Cindy, will come to say good-bye. She was like a daughter in the house, Mrs. Duncan has told me often. Cindy had her baby a week ago and she is going home the day after tomorrow. I know I can't measure up to Cindy here, but then the Jean T. who is here is really no one. It doesn't matter what they think of her.

I brought a cup of tea with me to my room, though I shouldn't drink anything so close to bedtime. I know I'll get my punishment during the night—countless trips to the bathroom and sleepless hours tossing in bed, but right now I don't care. I can let my hair down just for a while, let those tears roll—stupid, hopeless tears flowing silently without any purpose. I can't stop them for they come from somewhere deep inside me where I'm not the master. It frightens me that I can build up my careful defenses and watch them be shattered from inside myself.

It is time for my nightly ritual. I cross out another day on my calendar and then count the days ahead. January, February and half of March. I can see all the little black crosses and all the days in one glance. But I sit here and stare at the calendar and time doesn't seem to move at all, even if I stare for an hour.

Monday the 7th

We had a good day. I think I will sleep tonight. Cindy was here and we made cookies, a golden mountain of cookies, filling the house with their fragrance. Peter watched over the record player and provided us with Christmas music. Our cookie mountain grew slowly because we had to taste every batch. Cindy looked slender and pretty—and a little tired. Her baby has already gone to his new home. She told me about him; he had dark hair and blue eyes and dimples.

"They told me not to hold him," she said. "But I don't want to run away from it. It was really a miracle, Jean. Nothing like it has ever happened to me." When Cindy leaves the Home tomorrow, another girl will be gone—and another child's life just begun.

It is such a great thing—the birth of a child, and I feel so small.

Thursday the 10th

A letter from my brother Tommy, and a twenty-dollar bill enclosed! I haven't been this rich for months. What a wonderful feeling. I should, of course, put the money away for the future. When I get out of this in March I won't be able to go to work right away. But I'd rather do something extravagant, something quite unnecessary. I'll have to think about it. I go in for my checkup Tuesday—I can shop then.

My brother writes poetically about love. He's met this absolutely wonderful out-of-this-world creature, he says. She's a teacher for the American children on the base. My brother says he doesn't mind our family being apart this Christmas, as long as he can be with her. The letter doesn't sound like my practical, down-to-earth, reliable little brother. I'm happy for him, of course. Just hope he doesn't get hurt too badly, but maybe he's more sensible than I ever was.

I remember the first time I thought I was in love—maybe it's really the only time. I was seventeen, and we spent a great deal of time walking through the park holding hands. I can't remember much of what we did or even what we said, only that we were either totally happy or totally in despair. There was never any in-between—or at least I can't remember.

How things have changed since then. Now I find these moments of quiet peace and joy, quiet happiness without the exhilaration I remember—but a happiness that doesn't go away with pain and hurt. I pray that I will never lose it again.

Tuesday the 15th

Today was checkup day again. I'm still holding my weight down and the doctor says I'm doing just fine. Christmas was everywhere in the Home. All this month the girls do special projects and have extra entertainment. Major Laski, the tall, skinny, Polish-born officer, was in the office when I arrived. Her friendly smile was a welcome change from Mrs. Warren's usual sour greeting.

"Are you coming in to spend Christmas with us, Jean?" the Major asked. I shook my head.

"No, I guess I'll stay in the hills," I said. I didn't tell her that I don't think I can stand to have a memory of Christmas in a home for unwed mothers haunt my Christmases in the future. To be alone in a small room with a candle is better.

"We'll miss you," said the Major. "Perhaps you can stay one evening for one of our programs instead." I said perhaps, and let it go at that. Tonight they are having a Christmas carol sing-along and Thursday they will hang greenery and lights all over the house and decorate the Christmas trees. They'll bake cookies Monday, so next Tuesday when I come in I'll have a treat in store.

I could hear a piano and someone singing "Silent Night." The door to the TV lounge was ajar and I looked in. I couldn't see

the girl at the piano but the singer was facing me. Her voice was
so clear and beautiful that I stood there without moving. She is a
new girl and looks about sixteen; she has braces on her teeth, and
short blond hair with bangs above large blue eyes. She wore a
pale blue smock and her hands were folded over her full tummy.
In her eyes was that faraway look of enchantment I've seen in
children's eyes every Christmas.

I went downstairs for a smoke before reporting on the hospital
floor. In the corridor the smell of turpentine and paint hit me,
and the smoking room chairs were piled up outside the door. I
thought to myself, that's great, the janitor must've softened with
the Christmas spirit. I stopped in the doorway. The old drab sanctuary had been painted a cool baby blue, and in the corner was a
girl, in jeans and a washed-out dungaree shirt, down on her
knees putting the finishing touches on the floor boards. She was
whistling while she worked—that old sea chantey, "What Will
We Do with a Drunken Sailor." When she came to the refrain
she started singing, moving her brush in time with the melody,
"Way, hay, there she rises, Way, hay, there she rises, Way, hay,
there she rises—Early in the morning." That finished off the last
of the floor boards and she sat back on her heels to look the job
over. I cleared my throat and she turned her head.

"Hi, there," she said. "You're just in time to help me move
the furniture back in." She jumped to her feet and I noticed that
her stomach was very large.

"I'm Peggy B.," she said. She lifted one of the heavy armchairs
with ease. "I'm due next month."

"I'm Jean T.," I said. "I'm just here for my checkup today—
I'm staying in a wage home. I like the color."

"Oh." Peggy wrinkled her nose, which was spattered with
baby-blue paint. "It fits the mood I'm trying to stay in—cool. I
spend a lot of time down here and I'm hoping the color will re-

mind me not to blow my stack quite so often." She laughed and tossed her short black hair.

"This is a crazy place, don't you know. I mean I thought the girls in here would be a bunch of rebels like me—you know, not caring about the way most people live. I mean they're all having babies and all. And they're not that way at all." Peggy lit a cigarette with slender paint-stained hands. "These kids are the worst conformists I've ever met! They're acting as if they're failures— as if they don't measure up. Me, I'm having a kid, it's a great part of life; then I take off again on the road."

"Is that where you come from, the road?" I said. I took hold of one end of the table; Peggy took the other end and we carried it to its place by the window.

"Well," said Peggy, "yes and no. I've got a home someplace but I've been on the go for the last three years, ever since I quit college."

"What were you taking?" I couldn't help being curious.

"Education and art." Peggy laughed again. "Believe it or not, I wanted to be a teacher, teach little kids in grade school about art. It was a great idea. Little kids are great artists before they get harnessed. I wanted to save some talents. Too bad it didn't work."

"What happened?"

"Those damn courses in education drove me out of my mind. Ever take any? There's nothing more boring, and I can't see that it's got anything to do with kids, real kids." The furniture was back in place. The rickety old chairs looked even shabbier now against the bright walls. Peggy shrugged. "Maybe some other crazy nut-head like me will come along and fix up the furniture," she said. "Not that you girls deserve it." She looked at me and her dark brown eyes sparkled. "Did you go to college?"

I nodded. "Yes, only I got pregnant during my summer vacation."

"Chicken!" Peggy spat out some tobacco from the loose end of her cigarette. "Didn't have the guts to just up and quit?" I had to laugh. Three months ago I would have stalked out in anger.

"That's about it, I guess," I said. "But you got pregnant for a reason, too, didn't you?"

"Sure," she agreed. "Honestly, I just thought I'd try it. I've tried just about everything else—waitress in Alaska, barrel-racing on the rodeo circuit. Last summer I worked a steamer on the Great Lakes."

"And how does it feel?" I said. "Is pregnancy all you thought it would be?"

Peggy's eyes darkened and she lit another cigarette before answering. "Nope," she said. "It ain't what I bargained for—there's more to it than I thought." She sat silently staring out the little window high on the wall.

"I think the difference is that I can't quit this if I don't like it," she said. "I can't just pack up and leave. I'm involved with all of me whether I want to be or not. I thought bearing a child was part of the experience of being alive, that it was something I could go through and then leave behind. I guess I hadn't really thought about the baby."

"That's honest enough," I said. "So where do you go from here?"

"I don't think I'm up to trying the 'mother-experience' yet," Peggy said, and smiled a little. "The baby goes out for adoption the way I always planned it, and I'll pack my bundle and jump a bus for the desert country. I'm going to spend a couple of months on the trail with a horse and some paints and a couple of canvases. I'm gonna have to do some thinking after this."

"Jean T. Jean T. You're wanted on the hospital floor." The voice over the intercom was Mrs. Warren's and she sounded plenty impatient.

"If I don't see you again—good luck," I said to Peggy, and

ran. I heard her start whistling before I reached the stairs: "What will we do with a drunken sailor." I hope I see her again next week.

Tuesday the 22nd

Only three days till Christmas, and the Home was decorated with lights and greenery everywhere. They even had Christmas music playing over the loudspeakers in every room. Mrs. Duncan brought me in before lunch and I ate at the Home. I had hoped to find Peggy and maybe talk her into going downtown with me after my checkup, but she wasn't in the dining room. A girl at my table, Cecilie—she was there when I first came to the Home—told me that Peggy left two days ago in a rage. It had happened during chapel. Lieutenant Stewart was holding the service, and while the girls sang "Throw out the Life line, Throw out the Life line," she threw Life Saver candy wrapped in paper with printed Bible verses on it. After the singing the Lieutenant asked each girl who had caught a Life Saver to stand up and read the Bible verse out loud. Peggy had caught one of the candies and refused to stand up and read.

"Why won't you stand up, Peggy?" asked the Lieutenant.

"Because we have freedom of religion in this country and I don't share your faith," said Peggy.

"You should have seen her," said Cecilie. "You know, many of the girls in here don't go along with the Salvation Army beliefs. But this is a private institution and we've got to go along with their rules—or leave."

"You may believe as you wish, Peggy," the Lieutenant said, "but as long as you stay under our roof you will follow our regulations."

"I'll sit quietly in your chapel," said Peggy, "but I don't have to participate in your service."

"Please do as you are told, Peggy," said Lieutenant Stewart. Peggy stood up.

"You leave me no choice," she said. "I can't compromise, so I shall leave."

"She walked down the aisle," said Cecilie, "and nobody said a word or did anything to stop her. When we got out of chapel she had left and we haven't seen her since."

The girls are going downtown tonight to see the Christmas lights. Cecilie said several of the girls from wage homes are going along but I told her I have other plans. Somehow the thought of fifty or sixty pregnant girls going in chartered buses to see the Christmas lights goes against my grain. I'd much rather spend my evenings alone in my little room anyway.

I had decided to spend my precious twenty-dollar bill on something as silly as a maternity dress for myself. I really shouldn't have done it. I mean, there's no one to dress up for and I'll have to leave the dress at the Home when I'm through. But I suddenly had this tremendous urge to look neat and well-dressed and—well, Christmasy. These washed-out hand-me-downs are so depressing. It's almost funny, but one of the things my parents used to complain about constantly was my lack of interest in how I looked. My favorite attire used to be washed-out jeans and denim shirts.

So I took a bus downtown from the Home and walked into a maternity shop, trying not to look self-conscious. There was only one customer ahead of me. She was trying on a very elegant-looking maternity cocktail dress and looked like one of those ads in a women's magazine (pretty as can be—mother soon to be). She was chatting with the saleslady about the baby and I wanted to say, "Hush, someone might hear you." When it was my turn the saleslady asked me how far along I am, and when I said six months she looked at me carefully and said, "You certainly carry it well. Is it your first, honey?" I nodded and she said, "I bet your husband is proud. Do you want a boy or a girl?"

"Oh, it really won't make a difference," I said.

"Why, of course not, honey. You'll be happy either way, won't you?" She smiled and chatted constantly while she picked dresses off the rack and held them up in front of me.

"You're too pale, honey—can't wear this blue one. How about green?" I nodded. Anything would be fine with me. I wanted to get out of there. The saleslady had the kind of smile I hate, the kind older women seem to lavish on young expectant mothers. As if they were saying, "How sweet. I can just see your little love nest and the baby's room all ready." She wrapped my dress and pushed a blank form across the counter.

"Won't you sign your name and address, honey, and have a chance on a baby crib? We always give one to a customer every month." For a second I panicked. I couldn't very well tell her I'm going to give away my baby (although it might have been worth it to see the look on her face). I wrote Martha's name and address. If she wins, the baby will enjoy it, I'm sure.

Wednesday the 23rd

Someone has said that the happiness you give to others no one can take away from you. I will make Martha and Don and little Kathy Carver very happy one day in March. Stated that way, it sounds so simple. Then why do I have to be so mixed up inside?

A pretty young actress was interviewed on television this afternoon. She sat in a swivel chair next to a huge Christmas tree, swinging her shapely legs this way and that. "The thing I love most in this world is to always be on the move," she said. "I love to go to a new place even if I like where I am. Oh, yes, I always come back to Paris and I always fall in love there, just like that!" She snapped her pretty fingers and laughed.

I've never been to Paris, but I know what she's talking about when she says she loves to move on. Except that now I'm beginning to recognize the feeling as more like a need to escape. It doesn't sound quite as nice. I haven't exactly been moving around from place to place as much as I've been skipping from person to person, interest to interest, project to project, idea to idea. I used to think vaguely of myself as being dedicated to mankind, wanting to make the world better. Yet I've never been able to settle down to something specific and definite for any length of time. So I've eagerly thrown myself into a new friendship or a new idea, and when the newness wore off, when the moment came when I would be expected to give of myself, commit myself, be responsible for my share—well, then I'd find some excuse to drop the whole thing. I did that with Gene, the father of my baby—I know it.

Later

Mrs. Duncan came into my room and said she had something to say to me. Something she'd thought about for a while. "I don't want you to ever have company again without asking me first," she said. She was talking about Dorothy. "You understand, don't you? After all . . ." she let the sentence hang. I understood perfectly. I wanted to pack my things and leave immediately—but where could I run to? Instead, I lowered my head and mumbled that I was sorry.

"It won't happen again," I said. I don't know why she waited this long to tell me, but I know what she's been worried about. To her I'm an unwed mother from a Salvation Army home, who won't even tell her my name. She has taken me into her home and she trusts me with her children (and her silverware). I can't let it bother me. I've got to stick it out here whatever happens.

Thursday the 24th

The house is quiet. It is nearly midnight—Christmas Eve. The radio is playing "Silent Night" and I'm going to let myself get sentimental. The day has been a rat race. The doorbell rang constantly for deliveries—late mail, mysterious packages to hide away. Peter and Lori have climbed everywhere trying to discover secret hiding places; they finally went to bed about an hour ago when their father said Santa Claus absolutely won't come to a house where the children haven't gone to sleep yet.

How different from the Christmas Eves I remember as a child. Santa Claus came to our house in person on Christmas Eve. But first we always trimmed the tree; the rest of the house had been decorated days ahead, but the tree was left till last. In front of the fireplace was always a large tray of Christmas cookies and hot punch. We sang the Christmas songs as we worked and Daddy was always off-key.

When the tree was ready and lit, we turned off all the other lights in the room. Daddy sat down in the large chair by the fireplace and we sat on the rug around him. Mother brought the family Bible and we all listened to Daddy read the story of the birth of Jesus. As a child I used to wonder why my father's eyes always got moist when he read the part about the multitude of angels praising God and saying "Glory to God in the highest, and on earth peace, good will toward men" (St. Luke 2:13-14). Before I go to sleep tonight, I will take out my Bible and read for myself the second chapter of Saint Luke, verses one through twenty. I did it this morning, and I know that I'll probably want to cry again.

My father always had to go out on an errand on Christmas Eve after reading for us. And it always happened that Santa Claus knocked at our door while Daddy was gone. He carried a sack over his shoulder and spoke with the trembling voice of an old man. He asked each of us if we had been good children, and he knew exactly when we hadn't. Mother always asked him to taste our cookies and he took some and put them in his big pocket. I always wanted him to eat them right away to see if he really liked them, but he never did. Then he took a small gift for each of us out of his sack and promised to come back during the night, when we would be fast asleep, to fill our stockings. My brother always assured him that we would be extra careful to put out the fire in the fireplace before we went to bed. "So you won't get burned, Santa Claus." And Santa Claus Ho—Ho-ed and said that he would have an extra good present for an extra good boy.

After Santa left, Daddy always came back and said how sad it was that he had missed meeting him again, but that maybe next year he would come while Daddy was home. It took several years before my brother and I caught on. Even after that Daddy kept on playing Santa Claus and we went along with him without saying that we knew.

Those were the good years, before I started to grow up and have a mind of my own. Maybe some of my ideas were wrong, but my parents just couldn't understand that I had to live my own way. That I had to question the ideas they said were the only right ones, that I had to make my own mistakes. But then maybe parents always try to keep their children from making mistakes and getting hurt. I hope that when we meet again we can talk a little more calmly about things.

Daddy is working on a new book this year in Europe. It is about Christian ethics and the new morality. Even the title used to make me furious when I first heard it. Now I hope he gets it finished. I'll read it all the way through before starting to yell about it.

It is past midnight. Merry Christmas, Jean. Merry Christmas, little unborn baby. Christmas is all about birth, all about the hope and promise that came into this world with a Child. Hope and promise come into this world with each new child. You are a small miracle, little one.

Saturday the 26th

Holidays are hard on my equilibrium. I can't keep my mood even. I'd just as soon forget it's Christmas, yet I get sentimental over a candle and Christmas mail from Mother and Dad. They wish me happiness, of course, and all the best in the new year.

So I went for a quick walk up the road to calm myself, but even the hills couldn't do it for me today. Sometimes I wish they wouldn't be so calm and so beautiful. Maybe if a violent thunderstorm tore through our valley I would get my mind off my own navel for a little while.

Later

Something always hits you when you're low. Mrs. Duncan marched right into my room without knocking and said, "Why don't you answer when you're called?"

"I didn't hear anybody."

"Nonsense," Mrs. Duncan said. "You couldn't have helped hearing. Why do you lie about it?" Josephine is here to do the ironing and she had apparently called me to help fold some sheets.

"I don't lie," I said to Mrs. Duncan, but I could see in her face that what I said didn't make any difference. I couldn't keep the tears from coming, and when Mrs. Duncan saw me crying her voice softened.

"Don't cry, Jean. I understand. I sometimes don't feel like helping anyone either," she said. "But I must insist that you be honest with me as long as you live with us." I wanted to lift my head and scream in her face. Tell her I don't care a damn about what she thinks and I'm leaving right now! She thought my tears meant I was sorry, and put her hand on my arm.

"We won't talk about it anymore—try to relax now." She left the room and I threw myself on the bed and cried long and hard. Leave here? What would be the use of that? I would be miserable in the Home, too, and I can't afford to run up too big a bill there. If I tell Mrs. Duncan about myself, she may be nicer during the six weeks I've got to stay here. But I just can't tell her; I've got to guard against slip-ups in the future.

She called me a liar. Sure I am. That's why it hurt to hear it from her. I lie to her, to my parents, to the whole world.

Thursday the 31st

New Year's Eve. I'm home alone with the children. The Duncans are at a party and I've been watching television, all those wrap-up programs commenting on national and international happenings during the year. It's funny how they sound so much alike from one year to the next—little wars started here and there, a race riot, a strike, a government toppled in South America. An old statesman or a famous writer dies and they call it the end of an era. Eras always end and new ones begin.

This year certainly marked the end of an era for me. Last New Year's Eve I was at a party. It was a large one at the studio of a painter who'd "made it." I mean his paintings are selling for more than beans. We were dancing and laughing, toasting the new year—this year. I don't think I had any feeling of impending disaster. Not in my wildest imagination could I have thought I'd celebrate the end of this year here, under an assumed name, bulky with a child I'll never even know.

Where will I be a year from tonight? That, of course, will depend upon what I make of the year ahead. I hadn't quite thought of the future that way before, that the future depended upon what I would do. I used to think circumstances had a lot more to do with it. When I first got pregnant I thought of myself as a victim of circumstances I hadn't been in control of. I used to ask myself those big questions—I guess we all do—like *who am I, why am I living*. I haven't found the answers, but I think I have discovered one thing this year: I am responsible for what I do. I believe God created us with free wills, gave us the opportunity of choice, and so I believe that I am responsible for what I do. It doesn't sound like so much, but it is really an awesome thing, I think.

I used to blame so much of what I did on my parents. They didn't understand me, I used to tell myself. I am beginning to realize just how absurd that is. My parents may never understand me very well. But still I am the only one responsible for my action and I must take the consequences.

It's a thought to start a brand-new year on—a brand-new era. I don't know what it is going to be like, but I'm glad it's here. Happy New Year!

JANUARY

Tuesday the 19th

I can't think of anything nice to say about January. I can't stand this month. It always depresses me and this year it is worse than ever. Maybe it's because it is the letdown after the holidays. Maybe it's because it's the month I usually break all the good resolutions I made New Year's Eve. Or maybe it's just because it's the darkest month of the year and there isn't even a hint of spring to liven it up. Whatever it is, I feel I'm going to be lucky if I live through it. The hills look drab. I'm barely on speaking terms with Mrs. Duncan.

I went in for my checkup today and physically I'm in top shape, the doctor says. I'm glad he can't read my mind. I didn't see many girls I used to know. A lot of them have left and new faces are everywhere.

I did see Joan C. She delivered around Christmastime but her parents didn't want to take her home then. Cecilie told me all about it at lunch. She is just about the only girl I talk to nowadays. She's been around longer than I have and she seems to know everything about everybody. I never did find out anything about her, though. Maybe I will when I go in to stay.

Anyway, Cecilie told me that Joan had tried to talk her parents into taking the baby home, but they refused. When Joan first delivered, she called her mother to tell her, and her mother said, "Keep the news of your bastard to yourself," and she hung up. Joan had turned to the girl in the next bed. "I've got to tell someone about my baby. I've got to tell someone . . ." Joan tried to get her social worker to talk to her parents, but nothing worked.

Joan first came to the home in September, before I came here to the hills. She had the bed in the corner, and I remember her sullen face. She never talked much to any of us and we didn't try to talk to her. She was sixteen, and had come from a juvenile home. Her parents brought her, without luggage, to the Home; they had told her they were taking her on a Sunday ride. And then they dumped her here and said she'd get a beating if she ever ran away.

She did run away, twice. She got her beating and she was returned; after that, she mostly just sat in her corner. Except once. It was Sunday, and at breakfast time Joan came downstairs with a smile on her face that made her look different, almost pretty. She had washed her hair and brushed it till it shone, and she was dressed to go out. Usually, on Sundays local girls go home on a visit or for a ride with their parents or friends. All day long, cars pull up in front of the Home to pick up or unload girls. That particular Sunday morning Joan told us all that her parents were coming to take her for a ride. She beamed with pride when she told it and we were all really glad for her.

When the first cars began to arrive, Joan ran to the window to look out. But it was never the car she was looking for. Half an hour passed—an hour—two—then lunchtime. Joan's parents still hadn't shown up and the smile on her face wasn't so bright anymore. But still she ran to the window when she heard a car stop. At last, when most of the other girls had returned, Joan gave up hoping. She came to stand by the window next to my bed. Her eyes were dark and full of tears. Her mouth was trembling when she said—to no one in particular, "It must be nice to have parents you can trust. . . ."

Today I saw Joan dressed to go home, and crying on a green couch in the hall. Her baby is in a foster home and that's where he will stay, because Joan didn't want to sign his release for adoption. No one can force her. Maybe she hopes that one day she can get him back.

There are thousands of babies like Joan's little boy, and most of them grow up in foster homes and institutions because their mothers never manage to realize their dream of getting their child back. It is sad and I wish something could be done about it. A law isn't the answer, but maybe girls in homes like ours could get the counseling and guidance they need to see that their own despair isn't eased by keeping a child from getting a good home.

Saturday the 23rd

I've got to force myself to write letters nowadays. I'm afraid Mother and Dad are going to read between the lines that I'm not exactly happy. How can I write without telling anything about where I am and what I'm doing? I'm following the weather reports on television so I can write about the weather on the other side of the country as if I were there. I write about books and about news events and things like that, but I can't write about the job I'm supposed to be working at. I guess I could make one up. I'm afraid it's going to be too easy to check up on if I do say something about a job.

I have a hard time putting anything down on paper, even this sort of thing. I get so sick and tired of seeing my own stupid, hopeless thoughts that I could throw the typewriter out the window.

Walking is just about the nicest thing I do. I walk so fast and so far that my legs and my muscles ache all over. It's good exercise and it makes me tired enough to sleep.

About three more weeks to go—then I have to move into the Home. I'm counting the days.

FEBRUARY

Friday the 19th

I'm back in the Home with less than a month to go before my due-date. Wow, I didn't think I was going to last this long at the Duncans'. The strain was awful the last couple of weeks. We didn't have any big blowups, and when I left we said polite goodbyes, but I used to get up every morning hoping I wouldn't lose my temper before the day ended. In my better moments I thought of it as an endurance game. Now that it is over I'm glad I didn't pack up and run earlier.

I saw Miss O'Connor, my social worker, today and she said, "I thought you might have come back earlier—you had some difficulties, didn't you?"

"It was all a misunderstanding," I said. "They expected me to confide in them and I didn't want to."

"You were only the second girl there," said Miss O'Connor, "and they had grown very fond of Cindy. They were hurt when you didn't get as close to them."

"It's all over now," I said. "Maybe the next girl will be more like Cindy."

Miss O'Connor looked at the cigarette she was getting ready to light. "There won't be a third girl," she said. "Not for a while anyway. Our policy is that girls do not have to confide in their wage-home parents, and if they expect them to do that the strain can be an unhealthy one, we think. You girls are already in a strained situation and extra pressure is something we are trying to help you avoid." Miss O'Connor smiled a little. "I have watched you every time you came in for your checkup. If you had shown signs of being upset I would have told you to move back here earlier."

"Upset!" I couldn't help laughing. "You didn't see me rant and rave at those hills. I was just too damn stubborn to admit I couldn't get along."

"Well, it's over," said Miss O'Connor, "and I've got a nice surprise for you. You are going to sleep in the five-bed dorm downstairs. You'll have some privacy there. The girls are older—with the exception of Lisa."

"Wonderful," I said. "Thanks a lot."

"By the way"—Miss O'Connor looked straight at me—"you are still going ahead with the adoption plans for the baby, aren't you?" I nodded and she smiled, relieved.

"Good. Any time you need to talk, you know I'm here."

It's good to be in the small dorm. I haven't met the other girls yet, but the bed by the window doesn't have any sheets so I guess it's mine. Anyway, I've unpacked my things. Half of the chest of drawers next to the bed is mine and there is room on top for a few of my books. I can lie down on my bed and see the sky between two large treetops. I stay awake much of the night now—I can look for my favorite stars.

Sunday the 21st

A cup of tea, a cigarette, on a red counter. Is it really four months since I was here last with Karen? Toni brought my teapot

and nodded with a big grin, but I don't suppose he remembers
me—or Karen. She looked at the calendar on the wall and could
see her due-date. I remember I envied her and wondered if the
day would ever come when I could see mine.

I had a letter from Karen before I left the Duncans', post-
marked Colorado. She wrote that she's been skiing and it's grand.
When we were here together Karen thought she'd never go down
the slopes again. There was no return address on the letter, and I
don't know Karen's real name, anyway.

I am here alone now. I can see my due-date on the wall and
my stomach is hard as a blown-up football against the counter.
March seventeenth isn't far away now.

I bought a small notebook on my way down here. It fits right
in my pocket. I won't be able to write my thoughts down on the
typewriter in the Home. That first afternoon, when I was writing
in my room, one of the girls from the big dorm down the hall
stuck her head in our door and said, "Hi, stranger." She came
all the way into the room, and when she saw the pile of papers
next to my typewriter I could see her face go rigid.

"You aren't writing a book, are you?"

I laughed and said, "I sure ain't," but her eyes didn't leave
those papers, and she said, "We're all in the same boat in here
and we sure don't want to get famous because of it. Leave me out,
will ya?"

"It's just a letter." I grabbed a letter from Patrick that happened to be on my bed. The girl flopped down on the bed next
to mine and looked at me carefully.

"Name is Katie S.," she said. "You sure write long letters."

"Sure," I said. "It's for my boyfriend." I put the bundle of
papers casually down in my drawer, and Katie smiled.

"Gosh, I don't see what there is to write about in this place,"
she said. "Lousy food, nothing to do except a lousy movie every

Friday night, and the chairs are too damn hard to sit on all the way through the movie, anyway."

"Where are you from?" I asked.

"Milwaukee, a long way from home, thank goodness. What about you?"

"Chicago."

The closeness of our two cities made me feel a bit uncomfortable, and I noticed that Katie had that careful look in her eyes again.

I'm holding my breath. For eight months I've been lucky. No one has found out about me. What if something goes wrong now? Mother has sounded a little funny in her letters lately, as if she wants to come back *before* June, when they are due to leave France. I've tried to sound as cheerful as I can in my letters and keep telling her to see as much as possible while she is there—not to cut the trip short. I shiver at the thought that something may go wrong. . . .

The girls in my dorm are nice. Jane S. is the senior among us. She looks about sixty, with gray hair in a neat bun at the nape of her neck, her blue eyes deep in their sockets. She looks gaunt and worn but Cecilie says she is only in her mid-forties. She doesn't say much and walks around with her nose in a book most of the time. Cecilie says she teaches college and is working on her doctor's degree. Her parents live near here and come to take her for a ride every Sunday. I saw them from a distance today—both white-haired and small. I can't imagine how Jane ended up in the Home.

Lisa is due any day now although no one knows for sure when. She tags along after anybody who smiles at her. I think she has pains, because she often curls up on her bed next to mine, without a sound—only her eyes are wide and dark with long suffering. I want to reach out and take her pain and her fear away, and I get hot with anger when I think of the jerk who forced his way into her childlike world.

The other two girls in our dorm are German immigrants and speak English in a heavy accent. They are together most of the time, speaking rapidly in their own language. Their side of the room is immaculate. They aren't due till June and will go to wage homes next week.

Monday the 22nd

The Ladies' Auxiliary had gotten us some tickets to the symphony and several girls went last night. The bus came to pick them up as I watched from the window in the lounge. I love to go to concerts but I couldn't make myself go yesterday. I don't want to remember that once I went with my belly full of a baby. So I took a shower, and knitted before going to bed. The radiators in the house usually make a terrible racket but last night they hardly made any noise at all.

I was assigned to my new duty today. I'm in my last month now and the duty is easy; I do light everyday cleaning in the staff residence. The five Salvation Army officers who run the Home live in a small bungalow across the park from the big house. I make up their beds, run the vacuum cleaner, dust and keep the three bathrooms clean. I go there right after breakfast when the officers have gone to their jobs. I am all alone in the house and I like to pretend it's my own home. I dust off the knickknacks and the books on the shelves, straighten the pillows in the cozy living room and look at the pictures of their families. But even if I work very slowly my job only takes about two hours. I was through at eleven o'clock today and there was nothing more to do except sit around or go for a walk. I think I'll probably go down to Toni's more than anywhere else. Toni knows me now;

he brings the pot of tea before I have time to say anything, and he leaves me alone in my corner with my notebook.

Tuesday the 23rd

The doctor says I may deliver in two weeks. I called to tell Martha and she said everything is all ready and waiting for the baby. Her voice bubbled with happiness over the phone. We've been alone in this together, the baby and I, for so long now that it doesn't seem real that he is wanted and expected by someone else. He is so big inside me that I can't move without being aware of him. A few weeks from now he is supposed to be none of my business!

I go for my daily walk, and when one of the girls says she'd like to go along, I find some excuse. I don't want to talk to anyone. I'm in my own world and people on the outside don't matter. It is roller-skating time and the first flowers of spring are coming out on the lawns.

I walked down the main business street and the school kids were on their way home. They swarmed in large colorful flocks, full skirts floating over young girls' legs—bright chatter and much laughter. The sounds blended with the singing of tires over asphalt and the birds' chittering above. The girls have painted their young lips with pale lipsticks; their eyes are large and dark with heavy mascara. But even the masks can't hide how young they are.

I stop to browse through the rack of secondhand books outside the candy store on the corner. An old man owns the store; he shuffles behind the counter in his felt slippers, and his soft white hair is long and curling slightly. Round glasses rest halfway down his long nose, and his rusty old voice blends with the shrill voices of little boys who are always in the store buying candy and secondhand comic books for two pennies apiece. I like to listen to their voices.

I walk down the sidewalk in step with the words running through my mind: it's almost over, it's almost over, it's almost over.

Wednesday the 24th

I'm alone in the smoking room. I get through with my work before the other girls and get here in time for a quiet smoke all by myself. The staff residence is so neat every time I get there in the morning that I wonder sometimes what the officers do when they are off duty. They don't make a mess, that's for sure.

We had pancakes for breakfast today and Tammy at my table ate ten! I think about my weight now, and the baby seems to take up so much room in my stomach that I can't eat very much anyway. Mrs. Coralis, the second cook, read the prayer this morning and all I could think of was the soft clicking sound of her false teeth slipping back in place after every few words.

The girls are coming; I hear them in the corridor. We all head for the sanctuary when our morning duties are over. I should go upstairs to read a book or write a letter, or maybe take a nap, but I know I won't. I'll sit around here instead, listening to the gossip, watching the girls play cards or embroider. Maybe somebody will go into labor, and we can watch for the doctor's car and bet on the time of delivery.

Sue delivered yesterday. She is our youngest; she was thirteen last month. Sue's mother has eleven other kids at home and she has told Sue that she can take her baby home. Some of the girls have been to see Sue in the mothers' ward and they say she gets to hold the baby.

We can't go out on a pass till after lunch and the time drags.

I can walk through the hall pretending I've got an errand somewhere but when I get there I don't know what to do and I turn around to go somewhere else. I see other girls do the same; we drift listlessly through the house.

I peeked into the pink dorm and Nancy was on her bed crying. When she saw me she waved me into the room.

"See," she sobbed, holding out two snapshots. "My family, aren't they wonderful?" I looked and saw a middle-aged chubby woman and a balding man sitting in overstuffed armchairs with a Christmas tree behind them. The other picture showed the family at the table for Christmas dinner, three gawky little girls and a fat little boy in a high chair. They didn't strike me as wonderful, but Nancy smiled proudly through tears and told me she hasn't seen them in five months.

I walked upstairs and had to stop on the landing to catch my breath. Peggy ran past me, taking two steps at a time. She is fourteen and looks as if she is having the time of her life, always laughing or singing the latest hit song. There are several very young ones here now—they don't seem to mind being here as much as the older ones do.

Martha called. They will pick me up after chapel Sunday and take me out for lunch. I should have bitten my tongue out rather than say yes, but I said it anyway, in spite of myself. Martha wants to give me the clothes the baby will wear when he goes home from the hospital, but I don't think I can stand that. I'm going to ask her to give the clothes to Dorothy when she gets here.

Thursday the 25th

The two German girls have left for their wage homes and a young Negro widow moved into one of the empty beds today. She has two little children at home; her husband died just three months ago. Carrie has a wonderful big laugh and she talks about her little ones at home with such joy. She is due in three weeks but doesn't even look very pregnant. She told us she kept her job as a clerk in a department store till last week. She is staying at the Home because she has decided to give this baby out for adoption.

"I don't want my babies at home to know anything about it," she said. "They don't even understand that I'm expecting. I told them I had to stay away on a job for a few weeks." Carrie's husband was a Korean veteran and she gets a small monthly pension from the government now.

"But it isn't enough to raise three kids on," she said. "Colored kids got to have a good education to get ahead. I reckon I can manage to put two through college—I can't do it with three."

Friday evening, the 26th

The Ladies' Auxiliary are here entertaining us tonight. I enjoy the cookies and punch they bring but I can't say I'm thrilled by their performance. Tonight a fat lady in a tight corset gave a song recital, accompanied by a skinny, efficient-looking young matron at the piano. The fat lady sang in a quivering soprano, twisting a lace handkerchief while she sang. It's the sort of thing they make funny scenes of in the movies, but it wasn't so funny with fifty or so pregnant girls sitting on rows of straight-backed chairs while a small group of well-dressed matrons sat in a corner whispering and stealing glances at us during the performance.

I couldn't stomach it any longer. After the punch and cookies break, I stole away and came down here to the smoking room where I can be alone with my notebook for a few minutes. That is, almost alone. Jane from my dorm is over in a corner with her nose in a book as usual; her gray head is bent and she didn't look up when I came in. I know she's got to stay away when the Ladies' Auxiliary comes. Some of the ladies live in her neighborhood.

I don't know why my stupid pride keeps me from being just plain grateful when kind ladies make cotton smocks for us to wear, and layettes for our babies, and curtains and bedspreads for our "home-in-hiding." They give us all the materials for our hobby room and workshop, too, and I know they must spend hours of dedicated work to help us. Do I have a right to begrudge them the sense of satisfaction they may receive from doing something like that? Couldn't it just as easily have been me—giving and enjoying it—instead of receiving and resenting it?

Maybe what I really resent has nothing to do with the kind ladies upstairs. Maybe I'm remembering the afternoons and evenings when my mother was too busy with her auxiliary work to be home with us.

Saturday morning, the 27th

I'm alone in the smoking room, and the morning is gray outside. Everyone else has special duties on Saturday—scrubbing floors and walls, windows and woodwork, pantry shelves and stoves. There is nothing extra for me to do.

So much happens so quickly in here. Lisa had her baby last night, and this morning Carrie went upstairs to the labor room.

No one noticed that Lisa didn't go to the lounge to hear the program last night. When I came to bed I saw her bulky form under the blankets on the bed next to mine and I thought she was asleep. I woke up at the sound of her whimpering. The moon shone right in our window and I could see Lisa rocking slowly back and forth under the blanket. Jane was awake, too. She sat straight up in her bed across the room and her face looked pasty-white in the moonlight.

"Lisa is in labor," she whispered. "I'm sure—and the poor thing doesn't know what's happening to her."

I jumped out of bed and tried to lift the blanket from Lisa's head, but she held on and the whimpering stopped. Jane was beside me, a clumsy pale figure with a soft voice.

"Are you sick, Lisa?" We could see the head nodding under the blanket.

"Does your stomach hurt?" Jane asked, and this time we heard the muffled "Yes" and a sob. Jane looked at me.

"We've got to get her upstairs," she whispered. She patted the rocking form under the blanket with a gentle hand.

"Listen to me, Lisa," she said. "I'm going to take you upstairs and the nurse will give you medicine to make your tummy feel better." Lisa grabbed better hold of the blanket and held it over her head.

"She's scared—maybe we better get the nurse down here," I whispered, but Lisa heard me and her body grew tense.

"Go away—go away. . . ." Her voice was lost in a moan.

"Get the nurse, Jean—quick," Jane said.

I ran up the stairs two steps at a time and didn't even feel my big stomach in the way. I found Nurse Simpson in the office.

"Poor thing," she said, and ran ahead of me downstairs. We found Jane sitting on Lisa's bed. She held Lisa's head in her lap and was rocking slowly back and forth, stroking the tangled blond hair. She looked up at us with a quiet smile.

"I'm sure she's in labor," Jane said. With Nurse Simpson's help she gently got Lisa up and out of bed. The three of them walked slowly up the stairs and into the labor room. I followed just in case they needed more help. Lisa wouldn't let go of Jane's hand, even when she was put to bed. Her eyes were wide with pain and fright.

"Don't leave me, please," she sobbed. Jane looked at Nurse Simpson, who nodded slightly.

"I won't leave you, baby." Jane smiled at Lisa, and Lisa sighed deeply. Nurse Simpson gave her an injection.

"This will help," she said, "till the doctor comes. He will give the order to put her under complete anesthesia. Stay here till then, Jane." I watched them from the doorway, Jane's gray head and strained white face bent over the young blond girl tossing on the pillow. I thought, why, Lisa could have been her daughter.

Jane came downstairs at daybreak, her face drawn from lack of sleep. She tiptoed to my bed.

"Lisa's asleep now," she said with an even, almost toneless voice. "She had a little girl." Jane sat down on my bed and together we watched the white mist of dawn shimmer among the trees in the park outside our window.

"Little Lisa." Jane said it with a sigh. "She won't know what happened to her. She will go to a foster home now and I hope they'll be good to her. You and I will go on living, building a future on what has been. But Lisa isn't very bright; she is only trusting. Let's hope she won't always be let down."

Mrs. Larsen, the housekeepr, interrupted my quiet time in the smoking room. "What you doin' here? Get out till someone's cleaned up this mess." She nodded angrily toward the ash trays running over with yesterday's filth. I wandered out into the corridor and saw her go muttering toward the laundry. She muttered just loud enough for me to hear—I'm sure she meant me to. "Don't do nothin' but sit around smokin' all day—don't feel bad about what they've done. . . ."

Two girls came whistling toward her carrying a basket full of dirty linen between them. "No whistlin'—get on with your work," snorted Mrs. Larsen as they passed her. Behind her back the girls giggled and formed their lips again in silent whistling.

Late afternoon

I had to get away for a few hours. I have a pass for the evening, and I'll just have a cup of coffee here at Toni's instead of supper. The intrusion on your privacy isn't the only bad thing about the Home. It's the fact that I just can't help getting emotionally involved with the other girls. Maybe I shouldn't say that is bad, but right now it's upsetting.

Now I'm all upset over Carrie, the Negro widow from our dorm, and I shouldn't be. I didn't see the doctor's car come to the Home this morning, but the nurse on duty stopped me when I came from the dining room after lunch and said that Carrie wanted to see me in the mothers' ward. Carrie was sitting up in her bed, her lunch tray untouched.

"You made it in time for lunch, didn't you?" I smiled at her but she didn't smile back. When she spoke her voice was so low I had to strain to catch the words.

"Will you do me a favor, Jean?"

"Sure," I said. "Anything you want."

Carrie's dark hands fingered the sheet on her bed.

"I want you to call my sister and tell her the baby was born dead."

"I'm sorry. I didn't know." I didn't know what to say.

"You don't understand," Carrie said. "The baby isn't dead. He's doing fine. I just don't want my sister to know he's alive. It's bad enough for me to know." Inside, I wanted to cry.

"What's her number?" I said. "I'll call from the booth downstairs. What else do you want me to say?"

"Tell her not to come and get me. I'll take a taxi home Wednesday. And tell her to kiss my babies for me."

"O.K." I turned to go and Carrie called after me, "You see, I can't lie about it to my sister."

I dialed the number and the voice at the other end was so much like Carrie's, so deep and warm, that I almost hung up again. But I introduced myself and told her what I had to say. I heard the faint "Oh, no!" then silence. When she talked again her voice was low and muffled by tears.

"Was it a boy?"

"Yes," I said.

"Thank you for telling me, Jean," she said. "I am glad Carrie has friends there. Tell her that I pray for her. Tell her the good Lord must have meant it this way. It will be easier for her—later."

"Carrie is fine," I said. "She had an easy time. But I don't think she wants to talk about it when she gets home."

"I know."

"She wants you to kiss her babies for her," I said, and heard that muffled sob again on the other end of the line.

"Oh, I will, honey. I will. Tell her they're so good—and waiting for their mama."

"I'll tell her," I said. "Good-bye."

"Good-bye," said Carrie's sister, "and God bless you."

I walked slowly up the stairs again and found Carrie resting with her dark head buried in the pillow. She heard me coming and looked up; there was pain in her eyes.

"Your sister says your babies are good and waiting for their mama," I said.

"My babies," she said, glowing as I remember she used to do when she first told us about them on the dorm. "You know, this new one looks so much like them. That makes it easier." I nodded —there was nothing I could say.

"I found such a good family for him," Carrie went on. "A colored teacher. His wife can't have no babies of her own. They'll love my baby and give him all the things I can't give." She was silent and watched the gardener rake leaves from the driveway outside the window. She spoke without turning her head and I saw her hand tense on the sheet.

"What did my sister say when you told her the baby is dead?"

"She said it must have been the Lord's will," I said, and saw her hand relax again.

I almost ran from the mothers' ward and downstairs to get my coat and get out of there. The crisp air cooled my spirit, and I stopped at the drugstore on the corner to buy a candy bar for Carrie. Thank God for Toni's, somewhere where I can sit in a corner and watch people who aren't pregnant, and listen to conversations that have nothing to do with the agonies we live with in the big house on the hill.

Late evening

I took the candy bar to Carrie when I came home. I had to sneak past the nurse on duty—it was past eight o'clock and visiting hours were over. Carrie looked much better. She was talking to little Sue and laughing when I got there.

"Thanks, Jean," she said. She split the candy bar with Sue. "At least it won't make me as fat as I was yesterday." I sat down on the chair next to her bed, and soon Carrie's face turned sober and she stopped kidding around.

"Promise me one thing, Jean," she said. "Don't pick up that baby of yours. I know what I'm talking about. I've had two of them before and I know what it does to you." Carrie's eyes were distant. "You look down at that little baby snuggled so cozy and safe against you with his eyes closed. Something happens to you right there and then. You can't help it. . . . I had to pinch myself to make sure it was really me lying there with my own baby." It hurt to listen to her. But I know that telling me about it was sort of a substitute for holding that third little one.

Sunday the 28th

Sunday is an empty day here, no duties. Many of the girls leave on day-long passes right after chapel. The rest of us wander through the nearly empty dorms, go for walks, watch television and have lousy cold cuts for meals. Today I'm not going to stick around. I almost wish I were. Don and Martha are going to pick me up to take me out for lunch.

Little Sue came to chapel this morning for the first time since she had her baby. She wore a pink sweater and a yellow skirt, and her brown face was streaked with tears. She sniffed her way through the service. The injustice of things made me want to go light a torch somewhere. We've all heard what happened. Sue's mother and her eleven kids are on relief and Sue came here from Juvenile Hall. When Sue's mother said she wanted to take the baby and Sue home and keep them both on relief, the matter went to the juvenile court. Sue and the baby were made wards of the court. Sue will go to a foster home when she leaves here and her baby will be placed in another home.

I agree that neither Sue nor her mother is fit to keep a baby and that the baby will be better off in another home. But how did it happen that Sue was allowed to hold that baby—was allowed to feed and dress him—for five days?

When she came here she was twelve. She was too young to attend the high-school classes but she loved to sew. She sat in the hobby room and made dolls' clothes while her body grew big under the bright cotton smock. Her case worker said her mother would let her take the baby home and so Sue began to make baby clothes instead of clothes for her doll. Soon her cardboard box was full of little blue, pink, yellow and white clothes, all a baby would need. Five days after the baby was born, the blue door to the nursery was closed and Sue stood there crying. "I want my baby—I want my baby." They haven't found a foster home for her yet and she will be moved downstairs to a dorm again—away from that blue door.

Sunday night

We drove around town, walked through a museum and had lunch at a restaurant with a view over the harbor. Martha's eyes were on me often. I think she wonders if I'll change my mind. "When it's all over, after those six months of probation—I know *we* won't change our minds—" she said, "we'll take out a great big bottle of champagne." I thought, for me, that bridge will be burned and I'll have myself a good long cry. It was difficult to talk to them. I like them more the more I see of them, and I don't want to know them well or to have a great deal to remember about them. I think that if we had met under different circumstances we might have become friends. I didn't take the clothes Martha wanted to give me. I told her Dorothy will pick them up when she gets here.

They dropped me off downtown, as I didn't want to come back here yet. I saw a movie instead, Doris Day in a comedy. It was romantic in a silly way, and I cried.

I walked home; there were stars overhead. Cars whizzed by with blinding headlights. The night was cold and I walked with my hands deep in the pockets of my trench coat. I picked out the North Star, Stella Polaris. That made me feel foolishly secure. I talked out loud to the star as I walked, and Martha and Don and the baby were pushed far out of my mind for a while.

MARCH

Monday the 1st

My month is here! I used to think it never would come. But it did and today doesn't feel any different from yesterday—yet it is. Next time a new month rolls around it will be April and the world—my world—will have changed. Some of the question marks whirling around in my head will be gone.

I signed out for supper and came down here to Toni's just to celebrate the occasion by myself. I've got the booth in the corner and a steaming pot of tea and my notebook. Two truck drivers are sitting at the counter and Toni is reading aloud from *No Time for Sergeants*. Most of the time Toni is laughing so hard he can't read and the truck drivers are howling. The café smells of French fries and hamburgers frying on the grill, and outside dusk is turning into dark. I feel safe here and very much at home.

"Want another tea bag—on the house?" Toni grins and peeks under the lid of my teapot.

"Thanks, I'd love some more tea." I smile back, and Toni turns to the two truck drivers.

"Meet Jean, fellas, she comes from the Salvation Army Home on the hill, and fills up on my tea and scribbles in her notebook every day."

The drivers nod and smile and say, "Hi, Jean. Nice meetin' you." There isn't a thing in their eyes or their faces to make me feel embarrassed. They look as if they know the score and accept it as something quite normal. It feels good to smile back instead of hiding.

"Tough luck," one of the drivers says. He is tall, his dark hair streaked with gray.

"Oh," I say, "it's almost over now."

"Jeez," says the other driver. "Anything like that happen to one of my daughters I'd skin the guy alive!"

"Well," I say, and I can't help smiling because it feels so good to talk about it this way without pretending or trying to defend myself, "I guess it was my fault too—it takes two, you know."

"Yeah," the tall driver says. "But you gotta take the consequences alone. I sure wouldn't let the guy get off that easy."

"If I ever see him again I'll tell him," I say. "Thanks for taking my side, anyway."

"Jean will make it O.K.," says Toni and pats my shoulder. "She knows what she's doing." Both drivers smile and nod. "Good luck, kid," they say, and I smile back, feeling suddenly strong enough to take on the whole wide world.

I don't have anything to cry about. I've got so many things waiting for me and these months have made me stronger. They had better!

We have two new girls in our dorm. Anne is from Minnesota and she has told her parents she is working in New York. She was in love with the father of her baby, she says. He had told her he was in love with her and she thought he wanted to marry her, but when she told him she was pregnant he said he was sorry. Two weeks later he married another girl.

Diane is in Lisa's bed, next to mine. She is thirty-four and acts as if she's on Cloud Nine. Her bed is cluttered with bits of colorful yarn and cloth and she sings while she makes rag dolls and stuffed animals for her three kids at home. Diane is a widow and she will marry the father of her new baby as soon as his divorce becomes final. She is staying here in the Home to save on the expenses and to save her boyfriend embarrassment. Her eyes and smiling face have a glow that I guess goes with being an expectant mother in love with and loved by the father of her baby. She doesn't say "my baby"; she says "our baby." In here, the difference is to touch and feel. Diane's happiness affects us all. Her voice and laughter draw some of the girls like a magnet. They come to watch her make Raggedy Anns and hear her tell about her kids and her "honey" who loves them all. The pictures of her children and her boyfriend are admired and talked about. Every evening at eight fifteen her boyfriend calls. He is on vacation now, Diane says, and is spending his three weeks with Diane's kids.

"My honey will be so proud of his first baby," she says, patting her stomach. "His first wife didn't want any kids." The girls who come daily to sit with Diane and talk about her family want to hear again and again about her love. It is almost as if they are trying to submerge their own misery in her joy, as if they hope something will rub off on them.

The girls who don't come avoid Diane and don't even want to talk about her. "Ah," they say, shrugging their shoulders, "she's a fool. She's probably making up the whole story."

I think it is because Diane seems so *alive* in here, where most of us are walking around pretending we are just dreaming or something. That's why she affects us all, and no one can ignore it. She's shaking up the finely-balanced equilibrium we're all trying to preserve.

Tuesday the 2nd

The day started badly. The weatherman had predicted cold weather for last night and the furnace had been turned up. The heat was awful and nobody slept much. The weather was mild and I heard the rain against the window. I tossed and turned because of a million little aches, wishing one of them would turn into a real pain.

About midnight I heard Jane get up and go downstairs. Her slippers always trail along the floor. She looks horrible now, her cheeks sunken and her eyes hollow from lack of sleep. For her this must be living hell. I see some of the fourteen- and fifteen-year-olds run laughing up the stairs as if they're living a great big game, and I feel like shaking them. But they are perhaps better off than Jane who never walks down the middle of the corridor; she walks next to the wall, looking as if she would like to shrink into it.

After checkup today (I'm still doing fine) Simpson took us on a guided tour of the Hospital. Six of us are due this month. I like Nurse Simpson, and today she was smiling as she answered all our questions. I always had a feeling her stern face was just a front to keep us in line. I hope she is on duty when I deliver. With her and with Doctor Norwad around, I don't think anything could be too hard.

Simpson took us through the labor room and the delivery room and told us there is nothing to worry about.

"You're not going to be sick," she said. "Childbirth is a very natural thing and the pain is just your body helping to bring out the baby. And remember, someone will be with you all the time," she said.

We stood in a semicircle around the delivery table, and she

showed us the stirrups where our legs will be fastened, and the incubator all ready for the baby. Over the table are strong lamps and a mirror. Simpson's face looked stern again when she told us that we can ask to be blindfolded if we don't want to see our baby. A couple of the girls smiled and looked relieved.

I haven't made up my mind about it yet. After all these months of waiting and soul-searching, I don't know if I want to ask to be blindfolded. I'm going to pretend to others that I've never had a baby, but I'm not going to try to pretend to myself. I *think* I want to see the baby.

Lisa was taken to a foster home this morning. Carrie is alone in the mothers' ward now. All the dorms are filled to capacity and we hear that girls are on waiting lists outside. Miss O'Connor told me there are girls in the Salvation Army shelters and in rented rooms waiting to get in. About half the girls in here are due in March and early April. The nurses say it is always this way—a mad rush in March and then again in October. Now count back nine months from those two peak seasons and you get to June and New Year's. That makes sense, I guess!

After lights out

I can scribble in my notebook by the light from the lamp outside our window. It's getting harder to sleep every night, now.

Someone brought a portable radio downstairs to the smoking room after supper and some of the girls danced in the corridor. Their shadows jerked and swayed on the wall, grotesque and shapeless. The smoke in the room was so thick it burned our eyes and throats; most of us chain-smoke anyway. The talk drifted to "how we got here."

"Wow," said Cleo, a tall Negro girl from the pink dorm.

"I got so drunk at a party I don't even remember when it happened. I wasn't even sober enough to enjoy it!"

"For me it was wonderful," said Nancy, who is eighteen and a freshman in college. "I knew what I was doing. There couldn't be anything between us—marriage, I mean—but I don't regret it." Her face looked dreamy under the curlers. "We had one wonderful week together—it was worth every bit of what I'm going through now."

Wednesday the 3rd before dawn

I am alone in the smoking room, gray pre-dawn outside. I was tossing in bed after midnight when Jane whispered across the room. "Let's go down for a smoke." The smoking room was dark. Jane sat in a dark corner and her voice reached me, even and low.

"I went to a teacher's convention last summer and things didn't turn out the way I had thought," she said. "I guess I was desperate and I didn't care what happened. I wasn't quite sober that night and the man was married." The glow from her cigarette showed her pale face and thin wisps of gray hair across her forehead. She laughed, a sad little laugh. "And to think I could have gone to the Far East that month instead. Not much to joke about, is it? You see, I don't really mind the pregnancy—I mean the discomfort and all—it's just that I'm sure this is my last chance and I wish so much that I could keep the baby."

I sat there feeling guilty for being young, for having another chance waiting. There was nothing for me to say. I don't think Jane really wanted me to say anything; I was just somebody nameless to talk to on a sleepless night. After a while she put out her cigarette and shuffled upstairs. I heard the slow steps moving up.

Afternoon

This morning I found a small cardboard box at the staff residence. It was empty, just the right size for packing the things I need when I go upstairs to deliver. Major Laski came while I was dusting the living room and I asked her if I could take the box.

"Why, certainly, Jean," she said and smiled. "It's almost your time now, isn't it?" I said yes and thanked her for the box. Later I took it to the dorm and packed my things in it. I've seen other girls do that before. They keep the box under the bed and it's almost a status symbol. It means we're ready to go. I guess it's like the young expectant mother on the outside who packs a suitcase to go to the hospital.

But I was caught. Mrs. Larsen, the housekeeper, peeked in the door and when she saw the box she came all the way into the room. She looked angry.

"What do you think you're doing?" she said.

"Packing—to go upstairs," I answered.

"You know you're not allowed to keep luggage in the dorm. What if all of you started keeping boxes under your beds!"

"I've seen others—" I said. "Please let me."

"Don't 'please' me!" Mrs. Larsen looked as if she was enjoying the scene.

"You can keep it in the locker room if you wish—or I'll throw it out."

"But the locker room is locked at night, when I may need it." I was almost choking with anger.

"That's your problem," said Mrs. Larsen and left, slamming the door. I threw my things in a heap on the bed and ran downstairs to throw the stupid box away myself. I was so mad my hand was shaking when I lit my first cigarette in the smoking room. It doesn't take much to throw me into a fit nowadays. How am I going to stand up when the real test comes?

Thursday the 4th

I couldn't sleep last night and wandered downstairs to the smoking room. It was dark, but I saw the bulky outline in the corner and heard the muffled sobs.

"Is that you, Jean?" I recognized Jane's voice and turned to leave her alone. "Don't leave. . . ." The words were like a cry and I sat down. I had never seen Jane lose her composure before. I've watched her—she's always carrying a book around and she reads to keep from breaking. At least that's what I think she does. She reads, I write—for the same reason.

"I want this baby—but I can't, I can't." Her thin body shook with sobs. Then she straightened up and the words came tumbling out.

"I'm going to keep it with me the seven days I'm here," she said. "No one can take that away from me!"

"It will hurt more later," I said. Jane must have told herself the same thing again and again.

"I don't care." Her voice was defiant. "I'll cram it all into those seven days. If it hurts later, at least I'll have had that." We sat in silence. I wondered how she got her personal life so mixed up that the only way out was a home for unwed mothers.

Do we ever get to be free to do what we want to?

I'm not worried about being a fabulous success anymore. All I want is an ability to get along with myself and the world. Stability. How I used to hate that word. Now I want it. Stability—continuity—stick-to-it-ness, not to give up.

Friday the 5th

A new girl who calls herself Christine moved into our dorm yesterday afternoon. Early this morning she had a bloody show and was moved upstairs to Isolation. She is only six months' pregnant. Everybody is concerned for her. It is funny; most of us are unhappy about being pregnant, but we are all concerned if it looks as if one of us may lose the baby. Christine is in her late twenties, and she has been living with her parents and supporting them for the last five years. She is a secretary.

"They think I don't want to live with them anymore," she said, showing us the picture of her parents, a white-haired elderly couple sitting in a porch swing outside a small frame house.

"I don't know how they'll get along without me." Christine was blinking away tears. "I just couldn't tell them the truth. They are too old to be hurt this way."

Saturday the 6th at Toni's

I discovered a new road today, up a narrow winding side street, away from the main street with its busy Saturday traffic. The street climbed a steep hill and little houses stood in terraced gardens against the moss-grown gray cliffs. The tall crowns of old trees shaded the street and, above, white clouds raced by on a blue spring sky.

I could smell leaves burning, and somewhere someone was sawing with a handsaw making a singing steady rhythm. A small boy raced down the street on a bicycle, throwing newspapers on front porches with the sure aim of long experience. I saw a tree house in a big oak and heard children laughing. The street narrowed to a gravel road leading through a canyon with tall gray boulders, and there was the sound of water running in the deep ravine next to the road. The road wound its way around a boulder and there just ahead was a wooden bridge across the little creek. It led to a park.

I hadn't expected to find anything like that so close to the heart of the city—the tall trees, the paths covered by last year's fallen leaves, the picnic tables deserted this early in spring, and everywhere birds chirping and flitting around. Near the edge of the creek was a family of five kids; Mom and Dad were busy watching a small boy in blue jeans getting ready to put up his red and white kite. What a great day for flying a kite!

What a great day for doing anything, for walking down a path covered with last fall's brown leaves, crisp and dry in the sun, soft under my old shoes. A great day for leaning against an old rugged tree trunk, watching puffy white clouds in a clear blue sky reflected in the water of a small stream.

A great day to be alive, with a little someone alive and kicking inside me.

Sunday the 7th

The Major spoke about hope in chapel this morning. "Be of good courage, and he shall strengthen your heart, all ye that hope in the Lord" (Psalms 31:24). The Major said that in her opinion the word "hopeless" should be taken out of our language—it is the worst word, she said.

Jody is two weeks overdue and someone always asks her why she is still around. "'Cause I like it here!" she says, her brown eyes flashing below straight dark bangs. "I don't want to leave you all alone in your misery!" But when no one is looking Jody sits quietly just staring ahead. She looks tired today.

Last night some of the girls talked about inducing labor. We just got the news from upstairs—Suzanne took castor oil and is in labor right now. Suzanne is sixteen years old, chubby, blond and always giggling. Who would have thought she'd have the guts to do it? She isn't due for another two weeks.

Later

I took a walk through quiet Sunday streets. It is mild and sunny today and windows were open. I could hear radios or television sets in the houses. Lawns look spring-green, and I saw fathers playing with their children.

All is quiet in the Home. Many of the girls are out on passes, the others sit around talking, knitting, smoking.

Evening

I walked again after supper. It's getting so I can't sit still anymore. The moon is up now, so new and shining. Before it is full my baby will be born. He is all ready inside me: fingernails and hair, small and red and wrinkled and safe. I hope he is all right— he has to be, the way he kicks around.

I wonder what you'll be like. Will you ever wonder what your real mother was like, and why she wasn't ready to take care of you herself?

Tuesday the 9th

My name was left off the list of girls who go for checkup today. It was a mistake but I feel disappointed. It's just a week till my official due-date and I guess I was expecting some sort of a miracle, like having the doctor tell me the baby is ready to be born and I better hurry to the delivery room.

Cecilie was told she will have a breech birth. I wish she'd go ahead and tell her medical student that she is expecting his baby. She says she doesn't want to because then he'd probably quit school to marry her and not go on to get his degree. They've been engaged to be married for five years, Cecilie says, and all this time she has helped pay for his schooling.

I can't understand how she's kept this from him, or why she wants to give the baby up for adoption. But then maybe the story about the medical student is a lie. We all have lies in here.

I get a letter a day from Dorothy now. She is wonderful. Her small talk keeps me going. My drawer is piled with unanswered letters; each day I feel as if I'm drifting further away from the world I used to know. I wonder what this place will seem like when I look back from the other side?

Jody hasn't delivered yet. And I am the next one due. Some of the girls have started betting on which one of us will go first and they look at us as if we're two of a kind. Sure, we are both going to have a baby and maybe our babies will be born at the same time, but we are worlds apart. Her thoughts and her pain will be hers alone. I will have mine. When we go upstairs the girls will say to each other, "How wonderful, I'm glad she's through."

They will come to see us in the mothers' ward and between them and us will be a gulf no words can bridge. On their side is the waiting, the laughing and joking and quarreling and crying, the fighting about turns in the shower, the being together, sharing each other's lives—a little.

On our side will be the aloneness. They will look at us with something like envy and pity. There is something new in the eyes of the mothers and the girls shrink from it. Mothers don't come back to the dorm. They are still in the house, but almost forgotten. They are the reminders that the day all of us wait for and fear will come.

Last night in the smoking room the girls talked about carrying weapons. Some of them have switchblade knives; they say they need them for protection. Some of them have spent time in Juvenile Hall. I wonder what the knives have protected them from? Not from the boys who made them pregnant, obviously.

One of the knife-carriers is Lizzie, a bright Negro girl. I like talking to her, and she asked me if I want to go home with her tomorrow. She lives on the other side of town and I am really not allowed to go that far away. But maybe I'll go anyway. If I start in labor I can always get a cab—babies have been born in cabs before.

Wednesday evening, the 10th

We took a taxi across town, Lizzie and I. Riding in the taxi made me feel almost normal and the town looked just as it did when I saw it a month ago. Somehow I've felt as if the whole world has been changing with me.

Lizzie lives with her mother and grandmother in an apartment right above the Midnite Bar. Her father died in action in Korea. The street is wide and the old buildings are large. Once this was a refined residential neighborhood. Now the plaster is falling from the walls and the junk stores, bars, pool halls and grocery-liquor stores take up the first floors of the buildings.

The smell of stewing chicken met us in the steep hallway, where a lonely electric bulb threw a weak yellow shine over walls scribbled with names and dates and four-letter words. Lizzie's mother and grandmother met us in the kitchen, where the pot of chicken simmered on a stove in the middle of the room. The mother and the grandmother were both tiny and shrunken; Lizzie towered over them and they clung to her, crying.

"Meet Jean, from the Home." Lizzie introduced me and each shook my hand, showing a quick smile. In the living room plaster cupids crumbled from the tall ceiling. Lizzie put a stack of records on the corner hi-fi set, and just then we could hear the rumbling of many feet coming up the stairs. Kids started filing in and suddenly the room seemed full of them, laughing and talking, slapping Lizzie's back and glancing at me—white and very pregnant—on the edge of a chair.

"Jean is a friend of mine from the Home." Lizzie laughed and flung her arms wide. "Gee, but it's great to be out of that joint! Let's dance." Some of the kids nodded and smiled at me, but soon I was forgotten. Lizzie was dancing in the middle of the floor; it was obvious that she is the leader. In a loose sweat shirt over a tight skirt she didn't look pregnant. Her brown eyes were sparkling and her teeth very bright in a wide, happy smile. The whole room seemed to sway with the music—Lizzie was home again.

The door to the kitchen opened and grandma peeked in, beckoning to me. I went to her room on the other side of the kitchen. Here the dark red and white wallpaper had been left intact from former days. The lace curtains were pulled to hide the view of the scarred buildings on the other side of an alley. A bird cage hung by the window and on the dresser were rows of family pictures. The large four-poster bed was covered by a beautiful hand-embroidered quilt. The sound and beat of the dancing reached us faintly through closed doors but in grandma's room was an aura of timelessness, far from the rock-'n'-roll age. The room smelled faintly of something I couldn't recognize, except that it made me think of fields of poppies in spring.

Grandma pulled her shawl closer, and sat down in her red

velvet platform rocker. I sat on the edge of the only other chair in the room, a straight-backed wooden yellow one, and wondered what the shrunken old woman with the bright peering eyes wanted to talk about.

"We miss her so," she said finally, meaning Lizzie. "She is such a lively one." I nodded and waited.

"She is such a bright one, too, you know. First in class in school even if she never studies. She's too restless—just like her Daddy used to be."

"She'll grow out of it," I said, and grandma nodded.

"If she goes to college she'll get a pension," grandma went on, "since her Daddy died in the war." Lizzie's mother stuck her head in the door.

"Dinner's ready," she said. Grandma nodded again, staring straight ahead as she rocked back and forth.

"She's not a bad girl, just young."

"We all like her at the home," I said. "She'll be all right, I'm sure." I wondered if that was what grandma wanted to hear.

The kitchen was full of people talking and laughing and eating. The pot held enough chicken and dumplings for everyone, and it was delicious. I hadn't eaten like that for weeks.

After dinner Lizzie showed me her room and said I could take a nap while she went with the gang to the drugstore. She made me feel very old and tired, but I guess I wouldn't have been able to keep up with the rest of them for long. I lay on the big four-poster bed and stared at the books filling the shelves on the walls. Shakespeare, Plato, Nietzsche, James Baldwin, anthologies of Negro literature, pocketbooks on crime and passion, poetry—volumes of poetry.

On the dresser was a large notebook. The cover was scribbled full of boys' names and drawings of faces and trees. I opened it, feeling guilty, but too curious to leave it alone. Across the top of a page Lizzie had written:

Elizabeth—picture of a young woman . . .

Yesterday a girl—strange—longing, mind full of dreams. To run up a grass-clad hill—sing from the top.

Laugh in the face of the world.

Lonely, nights of wondering.

Different. Not just drifting along. Cutting off—cutting out—to be herself.

Seventeen—black skin—black hand—white skin—white hand—heartbeats in rhythm together.

The world is mine also. I want it.

Today—a woman. Woman's body. Mind—bright, alert, aware, knowing. And because of the knowing—crying.

Elizabeth—a girl's dream—a woman's secret.

Cry, Elizabeth. . . .

Last night in the smoking room she talked about carrying a knife and almost getting hooked for dope-peddling. Today here is her room with shelves full of books and her words, "The world is mine—I want it!" And a baby. What will you do, Lizzie? With your bright and restless mind—will you be bad? Or will you be very, very good? In the next few months you may make the choice.

One of Lizzie's friends drove us back to the Home.

"He's crazy," Lizzie said later. "He isn't the father of my baby but he wants to marry me anyway." She laughed. "But who wants to be married? Not me."

Later, after chapel

The Major spoke about Hidden Sins in chapel tonight. It almost made me laugh out loud.

They tell me Tina took castor oil and went into labor this afternoon. She is only fourteen and has been held in Isolation for two weeks for stealing from the girls in the dorm.

I haven't had a letter from Mother and Dad in nearly three weeks. Haven't written either, for that matter. I hope everything is all right with them and that they don't think there is anything wrong with me. My letters haven't been too cheerful lately.

Latest flash from the hospital floor—Tina's labor was false. "Serves her right," says Tammy, and puffs big clouds of smoke. Tina took twelve dollars from Tammy and has already spent it on a movie and candy.

I have piercing dropping pains and my stomach is hard as a wooden ball.

Thursday the 11th

Several months ago I dreamed one night that my baby would be born on March eleventh. I saw the date clearly in my dream and woke up thinking that it would be true, but I guess dreams don't come true often.

At breakfast Lieutenant Stewart led us in prayer. She asked the Lord to help us not to burden today with the sorrows of yesterday and the problems of tomorrow. I wish I had her faith.

The word went around at the table—Carolyn is in labor five weeks early and the doctor says it is twins.

There was a new girl at my table at breakfast. Her name is June and she is pretty in an all-American-girl-next-door way. Afterward, in the smoking room, she chain-smoked and talked very fast and very much. "Maybe this is tough," she said. "But it will sure make a girl think. There can't be much after this that'll scare me." Later she talked about love and said, "If we had loved the guy with the right kind of love none of us would be here. Maybe we'll be better judges next time." June is a psychology major from a Midwestern university.

Later, at Toni's

I've checked out for supper. Toni is boiling an egg for my supper and the pot of tea is on the red counter before me. This place is saving my life.

Tammy came back from town all upset this afternoon. She had gone down to celebrate her birthday; she's eighteen.

"My God," she said, pacing the narrow floor in the smoking room. "I ran into the bastard who's the father of my baby. I told him we was quit and the idiot says he's got a Reno divorce and wants me to move in with him again." Tammy flipped the ashes from her cigarette on the floor. "But I told him, 'You got no divorce when you should have so I don't need you no more.' And do you know what that self-lovin' fool had the gall to tell me? 'Sure Baby,' he says, 'Soon's I get you in bed with me you'll wanna shack up again.' " Tammy shook her head. "I hate his guts!

"So I fell for him once," she went on after a while. "So I was mad at Mom and had to get away. That bastard came through town and I went north with him to a lumber camp. When we come back my mom says to him, 'Get the hell outa here and don't come back till you're single,' but he didn't get no divorce then and I'm glad, 'cause he's rotten. He gave me fifty bucks today and that's all I'll ever get outa him, that louse!"

Tammy talks tough but I like her. She's always the first one to volunteer when a girl is sick and needs someone to do her work. Tammy doesn't have an ounce of self-pity (some of the girls do) and I think she'll go on from here walking straight and with her head up high. Red hair and laughing green eyes and a way about her that makes people turn to watch, even when she's pregnant.

Another cup of tea, a cigarette, and I'll stare a hole in the calendar on the wall. Every night I hope I'll wake up and have to go upstairs. When I have little pains I wish they would grow stronger. When I have an occasional strong pain I'm not quite so brave. But I wish I could get it over with.

Carolyn's labor stopped and they moved her into Isolation. She'll have to stay in bed for a while, the doctor says. But Carolyn is happy anyway. Her boyfriend is coming home from the Army in June and they'll be married. He has been paying her expenses at the Home, and when she leaves here Carolyn will take an apartment with her babies and wait for him.

"I can't go home," she says. "My foster parents and his foster parents won't let us have kids now—they'll try to make us give 'em up." Carolyn has been sewing on the layette for her baby since she came. Now, with twins coming, she is sewing in bed, they say. Her boyfriend is a mechanic and Carolyn has had a job since she was sixteen. She is nineteen now.

"I've known Don since I was thirteen," she told us once in the smoking room. "He was fifteen then and when I got to be seventeen I got pregnant. Don had just joined the Army and didn't know about it, and I was scared to tell him. My foster parents made me sign an agreement to give the baby up for adoption. They were gonna get a thousand dollars for it and they said they could sure use the money. But I lost the baby when I was five months along. My foster parents were mad but I was sure glad it happened. Nobody can take our babies from us now. We won't have a fancy home, but it'll be nice and solid." Carolyn's had the hard breaks but she's coming out ahead.

Miss O'Connor told me eleven girls are waiting to get in the Home now. The mothers' wards have empty beds waiting and we're crowded in the dorms. The bets are going strong about Jody and me—sure, we'll hurry!

The sunset is red over the dark rooftops of the city. I've spent two hours at Toni's counter; I don't know how many cups of tea I've had.

Two college boys are in for coffee. Their books are stacked on the counter and they wave their hands as they talk eagerly. Next year I'll be back in school. I'll talk with boys like those about art and philosophy and current events.

All this will be like a dream, but I'll always know it was my world for a while. And that I shared it with others—sweet girls, ugly girls, kind ones and tough ones with knives hidden under their leather jackets. But in the house on the hill we were women all, fulfilling the biological function of our kind. The rest of our functions would have to wait.

Friday the 12th

I wandered through the corridor, walked in and out of the dorm, picking up a book I ought to have read, putting it down again to wander around some more. Finally I checked out and came down to Toni's.

"You still around, kid?" Toni said, handing me a cup of tea before I had time to ask for it.

"I've got five days till my due-date," I said.

My stomach is hard against the counter and I watch two young couples in for quick hamburgers and Cokes on a Friday afternoon. The smoke curls up from my cigarette; I have nothing to do but watch it. When I curl my bare feet in my shoes I can feel a nail right under my left big toe—my only pair of flats and they're almost worn through. I can't afford to buy new ones. I can hardly afford to buy my little ten- and fifteen-cent notebooks. To make them last longer I write small and don't use a margin.

I shouldn't skip supper so often. Maybe I better go back for it tonight. But we can't sign out for a walk after supper, so if I stay in to eat I've got to stay there all evening.

I know what I'll do first when I get out of here. I'll walk all over the city one early morning. Climb the hills and watch the harbor, the fishing boats coming in from the sea, the big ships at anchor and at dockside, feel that fresh salt morning breeze come in from the ocean.

Later

Four of us were talking in one corner of the smoking room after supper. In the other end the regular poker gang was playing.

"Is this the way it feels when you're on dope?" said June. "Your senses so dull you only exist from day to day, happy you're not going mad?"

"And yet if you believe life has a meaning you've got to believe we're doing something important, waiting for new life to be born through our bodies," said Cecilie.

"Maybe I believe there's a meaning," said June, "but I sure don't know what it is—why we are born, live and learn through hard knocks, just to die one day. If God didn't mean for us to know why, why the heck did he make us bright enough to worry about it?"

I said, "Maybe at one time in our lives we fulfill whatever it is we are meant to do, only we don't know it."

"Whatever it is," said Cecilie, "I think a woman gets close to it when she gives birth to a child."

"That's something to be grateful for, anyway," said June. Madeline, who had been knitting all the time, looked up and laughed. "You are naïve fools," she said. "I've had three children and there is nothing beautiful or meaningful about it. Not even when you're married. It is ghastly and disgusting—the swollen bodies and the pain and all the blood." Madeline looked at each of us in turn and then went back to her knitting. None of us found anything to say in answer.

The doctor's car just pulled up. That settles the bets. Jody went to the labor room two hours ago. She certainly isn't wasting her time.

Getting ready for bed is the best part of the day. Everyone feels better at night. We've all put little marks on our calendars and counted one more day gone. Then to bed, stretching under the covers, searching for a comfortable position, not so easy to find now. But it is good to lie there, the restlessness of the day gone.

The window is open every night now, and the breeze is fresh and fragrant. I shift around and feel my stomach harden in contraction and wonder if I shall have pains before morning. But always the gray dawn comes and another long day of idle waiting ahead.

Saturday the 13th

I sneaked in to weigh myself in the nurse's office today. I've lost two and a half pounds this week—a good sign, say the girls who've had babies before.

There is a new girl in the smoking room. She sits fiddling with a package of cigarettes and hasn't even taken off her coat. Her short blond hair needs combing.

"How can you be so cheerful?" She interrupts the girls who've been laughing and talking as usual over their cards. "How can you joke about the babies?" Some of the girls stop talking and look at her. She goes right on. "It's cut-and-dried, isn't it? We're all in here to have babies we don't want. We're hiding it from the world and we'll leave here pretending it didn't happen. I hate those lies—and you just laugh!"

Tammy speaks first, and she smiles. "Cool it, kid—you'll get the hang of it soon." The new girl gets up to leave without a word. I see the tears flooding her eyes. By the door, June reaches out to touch her arm.

"We know we're in the same boat," she says soothingly. "We all cry sometimes and I don't think anyone likes to lie, but the tears don't help and we laugh to keep them away." The new girl looks straight at June.

"Maybe you don't like the lies," she says slowly, "but what about those poker players in the corner? Thanks for trying anyway." She removes her arm from June's hand and is gone. I feel sorry for her. I felt that way once and the adjustment hurts. I wish we could get more help with our adjustment in here. The social workers are mostly concerned with arrangements for adoption. We need someone who can help us understand ourselves better, and why we got into this mess in the first place. What we really need is a psychiatrist or a psychologist who could give us individual or group therapy. Since we have all demonstrated our problem in pretty much the same way, group therapy might be very effective here.

Sunday the 14th

Stella, the new girl, went into labor in a state of shock and was rushed upstairs right after chapel today. She was almost "discovered" in the kind of freak happening all of us are scared to death of.

Stella is a nurse and she had told her family she has a job here. Her sister happens to be visiting in town and phoned here yesterday. Stella talked to her, said she would be on duty all weekend and couldn't meet her. But her sister decided to come anyway. She came just as the girls were leaving chapel after the service this morning, and someone had left the front door open by mistake. Stella's sister was standing in the corridor. The Major saw the stranger and had the sense to pull her out of the hall and close the door before Stella walked by and the sisters didn't see each other. The Major explained that Stella was working in the lab and couldn't be disturbed, and the sister left without suspecting that Stella is one of us girls.

The front door was locked again and the Major called Stella to tell her what had happened. She came back downstairs to the smoking room and sat here pale and shaking. And then the pains came. Now she is upstairs in Isolation. They've given her sedatives to stop the labor, and she is safe from visitors. But we are all a little shaken. It could happen to anyone.

Monday the 15th

Last night the moon was almost full. I remember my first night in this dorm—a full-moon ago. I thought that by this full moon my baby would be born. I woke up in the middle of the night and the fog was thick against the window. I sat on a wooden stool in the bathroom and read about the true love of Elizabeth Taylor just to keep from having to go to bed.

The fog is still here this morning; we can't see the city from our hilltop. I walked across the park to my morning duty in the staff residence and thought how lucky I am. I had a letter from Mother Saturday and she says she will come home in May. That suits me fine. I'll have time to get organized after I leave here.

June just came into the smoking room, mad and ready to yell at someone. "I met that snotty Mrs. Horner—the group social worker—in the hall upstairs and she wanted me to take up a hobby. I said I didn't want to fool with it, and she told me I'm no better than the rest of them or I wouldn't be here." June lit a cigarette with a shaking hand. "Who the hell does she think she is, anyway?"

Later

The first session of the girls' council met this afternoon. There are two representatives from each of the three large dorms, one from our little one. I wanted to go but girls who are in their last month can't be members of the council. The council doesn't have any power other than to air grievances and make suggestions, but it is still a good idea. A Salvation Army staff member and one of the social workers are standing members. The girls' representatives are elected from the dorms, and the council will meet once a week. I think maybe the housekeeping staff ought to be included also since most of the problems come from our frictions with them.

Diane was representative from our dorm and said that the first meeting didn't accomplish much. They appointed a food committee to suggest more appetizing menus and avoid waste. Someone also brought up the point that certain girls make a practice of standing outside the door to the delivery room listening during deliveries. All the dorm representatives brought complaints about the heating system; it always gets too hot or too cold at night.

Tuesday the 16th

One day till my due-date. I went to the clinic this morning hoping the doctor would tell me it's coming, but he just said the baby is still floating, is not even engaged, and I was so upset I started crying later in the smoking room. Diane happened to come by with an armload of colorful yarn the cooks had given her and asked me what was wrong. I lost my temper and screamed at her, "Nothing's wrong! Can't you see I'm happy?" She didn't look hurt, just went on her way. I'm glad I didn't run into someone I don't like—I might have hit her!

It's drizzling rain outside. Just the kind of weather to suit my mood, but I can't get a pass till after lunch.

Later

Greasy spaghetti for lunch. I drank a glass of skim milk instead and then felt tired enough to lie down on my bed. I fell asleep and when I woke up my eyes were swollen from crying. But the sun was shining.

Cecilie wanted to go out with me, and we walked in the warm breeze with our coats open. I felt like smiling, and then I saw my reflection in a shop window, pale drawn face, dark circles around puffed-up eyes. I look awful!

After the second cup of tea at Toni's Cecilie said, "I worry about you, Jean."

"What do you mean? You don't even know me."

"Oh, yes, I do." Cecilie looked straight at me and I felt my stomach tighten—could she possibly?

Cecilie smiled and shook her head. "Don't worry," she said. "I don't know your real name and I don't care. But we've been here together now, since October, and I think I know a little bit about what you're like."

"Oh," I said, and wondered what she was talking about.

"You're a smart girl," she continued. "I bet you're filling those little notebooks of yours with thoughts about what is happening to you and why." She was stirring her coffee and didn't look at me. "I know I'm talking about something that's taboo around the Home."

"That's O.K." I wondered what she was aiming at.

"When you leave, don't try to forget." Cecilie's voice was urgent. "Don't try to 'make up for it.' Go someplace where you can be alone and think, and listen to what's inside you. Accept what is you and what has happened. If you don't, you'll be running away and you'll just run in a circle right back to where you started from."

"Get pregnant again?" I had to smile, it sounded so silly.

"Yes, that has happened to girls before." Cecilie lit another cigarette and we both stared in silence at the red counter.

"I'm going to break a promise," Cecilie said, not looking at me. "I think maybe you ought to know."

"Go on," I said.

"Jane in your dorm had another baby eleven years ago—here in the Home. She gave it away for adoption without even seeing it."

"No!" I felt as if I was going to be sick. I remembered Jane's drawn face, the thin wisps of gray hair and her voice: "For seven days I want that baby. I'll cram it all into those seven days. I won't have another chance."

"Oh, yes," said Cecilie. "She told me herself. She never did get over feeling guilty. Feeling like a failure. And she always kept looking for that baby . . . wondering every time she saw a child that age." I found it hard to look into Cecilie's face.

"Just being intelligent won't keep you from making the same mistake twice," Cecilie said. "You've got to learn to accept yourself and the world around you as they really are. Jane couldn't."

We walked home under a sky colored red by the sunset. Cecilie didn't say any more but my head was full of confusing thoughts.

I've thought so much up to the point when I will leave here, and then maybe about how I must keep from letting myself ever interfere with the baby and with Martha and Don. But somehow my thoughts haven't gone on much from there, to how I will live with myself and the world around me, what I will actually do. I've been vague about it. I've never thought that it might be too hard, that I couldn't somehow do it. To think that Jane, who has accomplished so much, couldn't!

Tuesday night

Jane and I were both moved upstairs to Isolation tonight, to give room to new girls. I share a room with Stella and Jane is alone next door. Nurse Simpson told me that Jane will be allowed to stay with her baby in that room, after she delivers.

Stella still looks pale and shaken after the near run-in with her sister. "I'm glad you're here," she said when I came in. "I can't stand these thoughts and the time dragging endlessly. Six more weeks to go. There are some people right now who know they only have six more weeks to live, six weeks to do what they want to do before they die . . . and we are just waiting, wasting our time." Stella has deep dark circles around her eyes and she's grown frightfully thin.

"It shouldn't be possible to just waste time," she said. "I can't help it. I can't help feeling how horrible it is to waste all this . . . and then not know where to go and what to do when it is over. I don't have any money to go home on. My family thinks I have a job and am well off. Even if I'm lucky, and strong enough, to get a job the day I leave here, what do I live on till my first payday comes around?" Stella hides her face in her hands and I can hear her sob. But what can I do about it? I'm in the same boat. I don't have a penny to share.

MARCH

Wednesday the 17th

I've circled today on my calendar and now that it is here it doesn't seem so important, my due-date. What will happen will happen in its own time. I can't push or postpone it.

Cecilie went along on my walk again this afternoon. "Can't let you go alone anymore," she said, and for a moment I didn't know what she was talking about. But of course—the baby may come any time now.

Toni greeted us and brought our coffee. "On the house today— your day," he said and winked at me. "Let me know when the baby gets here."

Tina left yesterday and it was discovered that she had stolen money from all the dorms. Not much, but then the girls don't have much. She took the six dollars Jody had saved for bus fare home, and Jody is due to leave tomorrow.

Sunday the 21st

It is almost midnight, but I can't sleep. I'm alone in the smoking room, and outside is the steady sound of rain beating against the ground.

I have a feeling deep inside as if something terrible is about to happen. I am scared. Is it the baby I'm scared of? I don't know, but I have never felt anything like this—not any pain, not anything physical, but as if I were trapped and helpless and something was closing in on me. I wish I could hide in a dark corner and give birth to my baby—just the two of us alone, no hands and thoughts of others intruding on our world. . . .

Monday the 22nd

This can't be true—it's like a nightmare! Mother is coming! She's arriving in New York tomorrow and thinks I'll be at Dorothy's to meet her the day after!

I cried hysterically after I read the letter, but at least I'm calm enough now to smoke and to write. Mother says she can't stay away any longer; she's flying home ahead of Daddy. "We'll have a wonderful week together," she writes, "and then I'll have to go back to New York to open up the house and get things in order for your father. Tell your wonderful friend Dorothy I'm looking forward to meeting my little girl's 'second mother.' "

What am I going to do? I'm stuck. I'm sitting here in the basement, smoking and staring at the pipes on the ceiling. My belly is swollen with an unborn baby. Why? Why? Why? I can accept the pregnancy. I needed to learn my lesson. But Mother! Must she know this? I've got to think of something. Ask somebody to help. To do what?

Later

I called Dorothy and made a mess of myself crying over the phone. She sounded so calm, so sure. She said not to worry, it will work out. She'll stall Mother—tell her I've gone skiing in New Hampshire or something, that I'm staying in a cabin with friends and can't be reached by phone. She'll tell Mother I'll come to New York to meet her there after I get back.

"That'll give you at least a week," Dorothy said. "Something will happen soon, don't worry. The baby can't wait much longer. It is overdue already." Maybe Dorothy is right, but on the other hand I've heard of babies being born several weeks late. Mother is bound to think something's up. And even if the baby is born right away, how can I face my mother seven days after I've had a baby and pretend I've been skiing in New Hampshire?

June is sitting here with me in the smoking room and says I ought to talk to the chief of the medical staff. He is here at a conference today. Maybe he could check me and see if I'm ready; maybe they can induce labor.

The word is out. Everyone knows by now and everyone is full of sympathy and good advice. I can see the fear in their faces. They're as scared of being discovered as I am. Mrs. Tebbits, the first cook, just brought me a cup of tea and said she is praying for me. Funny, but her clicking teeth didn't sound so bad, and I am really glad for her prayers.

Later

I've been upstairs and the chief of the medical staff will look me over in half an hour. Nurse Simpson told me to be ready. My stomach is hard as a rock. God, make me ready. Let that little baby be ready. We've gone through this much, we've got to make it through the last stretch now.

Later

Nurse Simpson stood by while the doctor examined me. He gave me a pelvic examination, and it hurt. Then he told me to sit up, and he patted my knee and looked at me with kind eyes. I heard his voice from far away.

"Sorry, young lady, I'd like to help but you aren't ready yet."

"How long?" I said.

"It's impossible to tell. The baby is still floating. It could happen anytime or maybe not for two weeks." Nurse Simpson took my hand and squeezed it tight.

"That's all right," I said. "I understand." But inside I wanted to scream and rage at the unfairness of it. I stood up and felt my knees shaking. I had to lean against the table and I wished I could just pass out, lose consciousness, lose myself. I remembered to thank the doctor for spending his time with me, and I walked without feeling my legs down the corridor to my room, where I got out a textbook I should have studied a long time ago. I took the textbook downstairs to the smoking room. The girls didn't have to ask me what had happened; it shows all over my face.

I took the cigarette June gave me and opened my book, an outline of American literature. I saw the letters and the words, and behind my eyes they collided with the thoughts that are tumbling around, making my head swim. It's no use. My stomach is hard again, and it is resting on my lap as if it were something separate from me. It is. I can't control it. I have no power over it. I am trapped by the inside of that stomach of mine.

Suppertime

I have only half-hour passes now, since it is past my due-date, but I don't give a damn. June walked with me to Toni's through the rain and I'm not going back to that place on the hill till it's bedtime.

Toni said, "Hi, you still around, kid?" when we got here, and I said yes and smiled automatically, as I've done every day this week. The Coke-calendar with my meaningless due-date hangs on the wall.

June spent her money on the juke box, and told me about the father of her baby. He is married and she is in love with him.

He wanted to get a divorce when he found out she was pregnant, but June's father threw him out of the house and told him never to come back. "So," said June, shrugging her shoulders, "this baby goes out for adoption because we weren't ready for it. Maybe some day he'll get a divorce and we'll meet again."

June has kitchen duty today and had to go back for supper. I took a walk by myself in the rain. My shoes got soaked and the rain seeped through my coat. The drops rolled down my face and neck and mixed with the tears I couldn't hold back any longer. I talked out loud to the rain and the trees and the passing cars. All will be well, I told myself, as long as I don't lose my head. I walked as quickly as I could, carrying my big stomach. Every step hurt but the pain was good to feel. The rain poured into little streams that ran along the sidewalk, and the lawns smelled fresh and clean. Dusk turned to dark and I kept on walking.

How did I ever get into this mess? I've had nine months of crazy luck, and now when it's almost over, the whole thing blows up. My thoughts went round and round.

Back at the red counter, Toni looked at me with serious brown eyes. "What's wrong, kid? Don't go sick on me in this weather." Toni is a great guy. I ordered a soft-boiled egg for supper. I've got to keep my health through all this or I'll never make it.

June is back again. She is all upset because the kitchen was a mess. She tried to hurry and spilled a whole bowl of gravy all over herself. The way she told it, it sounded so funny I had to laugh, and we both laughed till our stomachs cramped.

After lights out

The girls were all excited when we came back. Peggy had called her father, who is a doctor, and he told her that inducing labor at this stage in a pregnancy shouldn't present a problem. Everyone thinks I got a rotten deal from the chief. Another girl has a boyfriend who is a mate on a ship leaving for Hawaii in a couple of days.

"I can get you some postcards tomorrow and you can write them with different dates. My boyfriend will mail them from the Islands to your mother," she said. Most of the girls are in favor of that scheme and they were full of ideas about things I could write on the cards. Nobody talks about anything else and the excitement has everybody in a frenzy. All the girls can remember how others have been in tight squeezes before and they tell me and each other that no one has ever been discovered yet. We're all in this together, and the girls act as if somehow we can work together and find a way out, too. This feeling of solidarity in a crisis is real enough to touch. It has even spread to the staff. June told me that the Lieutenant prayed for me at the supper table.

Tonight I feel there must be a way. I don't know how, but I think all will be well. I am tired—tomorrow I will face what comes.

Tuesday the 23rd

I feel as if I'm floating in a happiness greater than anything I've ever known, beyond the range of words I know.

My baby was born four hours ago! A beautiful, perfect, live little boy, born through me. That such a miracle could ever happen to me!

People say newborn babies are ugly. Not my son. The nurse held him upside down, a bright red, naked little human being, screaming furiously. Who can blame him? He's been pushed and squeezed and handled, and then dangled upside down in a brightly lit, strange world.

How silly my fears have been. If I had only known how wonderful it would be to see him alive and perfect, to know that he has grown inside me all this time, I would have longed for it instead of dreading it. Even the lies, the loneliness and the hiding —today I feel they were worth it just to be alive to this moment. He is someone else's son, but to me was given the miracle of giving birth.

I feel as if all my life has only been a prelude to this, that I could die now, fulfilled.

Later

I called Dorothy and told her the good news. She didn't sound surprised at all. "I was awake most of the night," she said, "and toward morning I felt that something had happened. I knew all would be well."

Dorothy will stay home till Mother arrives to make sure she doesn't get worried about me. "I'll tell your mother you'll be coming to New York in ten days," she said. "I'll fly out to see you Saturday and we can go home by train to give you time to collect yourself."

Then I called Martha and Don to tell them they have a son. "He is perfect and very good-looking," I said. Inside I felt a swelling pride that almost choked me.

"A son!" said Martha and her voice rose with excitement. "How wonderful, how wonderful! I can't wait to get him home!"

"Next Tuesday, on our seventh day," I said, "he can go home." Not even the thought that he belongs to Martha can mar the joy of today. I won't let it. My body is so light, so relaxed. My stomach is so flat; it is funny not to have a little someone kicking around in there.

I am going to sleep now—on my stomach.

Just to think—he was inside me. . . .

Wednesday the 24th, First Day

They call the day the baby is born Zero Day, so this is our First Day. I slept most of the night but I am tired. I guess I was too excited to feel anything yesterday. The nurses are so sweet, so different now that I'm a mother. They are stricter with the girls. Grumpy would be a better description.

Jody and I are the only ones in the mothers' ward and we can hear the babies crying across the hall. I imagine I can recognize the cry of my little boy. We are being waited on like queens. Yesterday the cook even put a small vase of flowers on my tray.

June came by to say hello from Toni. She said the girls talk about my delivery as a "miraculous escape" from discovery.

"I guess from now on you will be one of those stories that will be told from girl to girl for years to come," said June. "Like the story about the girl who got out of here only two hours before her mother was due to arrive in the city." Unwed mothers must have a protective saint somewhere looking after them, I guess. Labor wasn't really hard, and not as long as I had feared. At first I couldn't even believe it was anything other than severe gas pains. Nurse Simpson was on duty and I shuffled down the hall to ask her for a pain pill. She took one long look at me and said, "You better go to the labor room and let me see how you're doing."

"But the chief said it might take two weeks," I protested. "I just have a few cramps, that's all."

"Babies have a way of ignoring doctors' predictions," said Nurse Simpson. "You better get in bed." She checked me and said I wasn't far along yet, but it was labor all right. I kept telling her she was probably wrong because this wasn't at all the way I had thought it would be.

"Won't hurt to get you ready," she said calmly, going about her preparations. She gave me an enema and I thought the baby was going to come popping out while I was sitting on the stool. But it didn't, and when I got back into bed Nurse Simpson told me to relax and to remember that the pains served a good purpose. Each of them would open up the birth channel a little more for the baby. Then she left me alone.

"Push the button when you need me—I'll be in the next room," she said. I had been afraid of labor mostly because it was something no one could tell me about. Now the pain was just something to go through. I could duck headlong into each new pain-wave knowing that it brought me closer to the moment of birth.

Sure it hurt—I can't even remember how much, except that each time it came I knew that I was made to take it. The pains came close, about three minutes apart, from the very beginning. I guess that's how come I thought they were only gas pains. I had heard so much about timing pains. Now I know there are exceptions to every rule. After just a couple of hours the pains got so bad I had to push that button. I was a little bit ashamed because the nurse had told me the baby wouldn't come for a long time yet.

But when Nurse Simpson examined me again she looked surprised. "Time to call the doctor," she said. "You've opened up quicker than I thought." I kept hoping it would be Dr. Norwad, and when I saw his eyes surrounded by dark circles above the surgical mask, I remember I thanked him for coming.

When it was all over I felt like laughing and dancing, and yelling from the rooftops for all the world to hear that I'm a mother. I remember I told Dr. Norwad I wanted to have lots and lots of children one day. He laughed and said, "Great. Just don't come back here to have them!" It sounded like a funny joke then.

Today I feel sore all over. My muscles are aching from the

strain they've gone through. The nurse says that's only natural, that the First Day is always worse than Zero Day.

Here is another new mother. It's Stella; she had a little girl five weeks early. Stella is sleeping; the nurse says the birth wasn't easy but the little girl is doing pretty well in the incubator.

Thursday the 25th, Second Day

Two new mothers today—I guess the ball is rolling and the late March rush is on.

I feel fine today. I weigh five pounds less than when I first got pregnant and my stomach isn't flabby. I think I can leave here and look pretty much my normal self.

Of course I have been putting off thinking about leaving. I don't know how I'll feel when I see Mother, but I know I can't give myself away, not now that it is all over. I must think of getting a job right away too, and get back to studying. I can't let this feeling of emptiness creep in—it is hovering on the edge of my mind now. I've got to wrap up this experience, know what it has meant and then go on my way. I needed these nine months. Now I've had them and I must go on.

There is a brand-new little boy and his family is waiting for him. He is no longer a concern of mine. I have served my purpose. So I couldn't be a mother for you, little boy, but you will have a mother. You will be loved very much. You have helped me, but someone else will meet your needs now.

Friday the 26th, Third Day

The doctor says I can leave next Tuesday. I'm in fine shape, he says. Dorothy will get here tomorrow and she'll call Martha then. They will make arrangements to meet Tuesday morning and Dorothy will bring the baby's clothes here. Then she will carry the baby out of the hospital to Martha. Dorothy will stay here and we will leave together Tuesday afternoon. It is all neatly planned.

I can get up and walk around now. When I had my shower yesterday it felt wonderful to be able to bend down and wash my feet without having to stretch around a big stomach. Tomorrow I will start taking my meals in the diet kitchen and I will have a job—rinsing out diapers before they go down the chute to the laundry.

Friday afternoon

One of the nurses came in to tell me that my little boy has been sick. He ran a slight temperature on his second day, yesterday.

"His temperature is normal now," said the nurse, "but this may mean that he can't go home Tuesday. He may have to be under observation a while longer." The nurse smiled and patted my arm. "Don't worry, hon," she said. "I just thought you'd like to know about it."

I haven't called Martha yet. He isn't sick anymore so maybe they'll let him go home Tuesday anyway. I can't believe that something will go wrong now. I walked across the hall and peeked in the door to the nursery. I saw that his crib has been moved behind a screen, separate from the others. I tiptoed in to look at him more closely. He was sleeping peacefully with one fist jammed into his mouth. Poor little one. I hope he isn't hurting. What would it mean for all of us if he is really sick?

I've got to get some sleep. Maybe tomorrow I can speak to the doctor.

Saturday the 27th, Our Fourth Day

Dorothy called. She is staying with her aunt; after all, that is the excuse she is using for coming here in the first place. She said my mother is fine and not at all worried about my being away for a couple of weeks.

"She said she knew you weren't expecting her home now, so how could you possibly have known," said Dorothy. I told her about the baby's temperature and she said I ought to call Martha.

"Don't get upset over it," said Dorothy. "It's probably nothing serious, but they ought to be told." I called Martha right away and heard the catch in her voice when she said, "Is he all right?"

"Don't worry," I said. "He doesn't have a temperature anymore, and he sleeps well and is gaining weight. It will be all right."

The pediatrician gets here Monday to check the babies and then he'll call Martha's doctor to talk it over with him. The nurse tells me he doesn't have a big appetite yet, but he is improving all the time.

Just two days ago I wrote in my notebook that he is no longer a concern of mine. Now I am being forced into caring much more than I had planned to. It will be harder to forget this way, if forgetting is what I want. But he is my concern until the moment I know he is safe in his mother's arms.

I have been to see him several times today. He looks so content and so healthy. His hair is a fine reddish brown and his eyes clear blue. There is a dimple in his chin. I am sure he'll be a real heartbreaker one day. I could stand by his crib just watching him forever. He wrinkles his nose and stretches his arms way up. Then his face gets all puckered up and he yawns or he cries, and I don't know what to do about it. Sometimes when he is asleep he mutters and moves his tongue as if he is searching for the nipple.

I am sure he smiles sometimes but the nurse says it is only gas pains. I pretend it is a real smile. I'll never see the way he'll smile some day. I must see in him today all he will be in the future. I feel so close to him now, in a different way, not just because he came through me, but he has become someone entirely his own. I can't let him down, for *his* sake.

What if Don and Martha won't take the chance? For a moment I feel a surge of excitement. He would be mine then. But of course they'll take the chance. And they have everything to offer him—I have nothing.

I'm going in to say goodnight to him now.

Sunday the 28th, Fifth Day

I watched the nurse feed him and change his diapers today. She cuddled him and talked to him and I was allowed to stand close. He is so beautiful.

You just have to be all right, little man. It is a little bit up to you, so drink all that good milk in your bottle. With a burp in the middle of your mealtime, you ought to make it.

Grown-up people will decide your future soon. Oh, but I know they will decide the right thing for you. Had you been born to them they wouldn't hesitate.

Monday the 29th, Our Sixth Day

The nurse told me to come to the doctor's office. "It's about your baby," she said, and the words shot through me, *your* baby.

He is my responsibility. The doctor said he can't release the baby for adoption now.

"I must be sure he is perfectly healthy before I can sign him out," he said, "and I can't be sure until we have observed him for at least several months. A fever may not mean anything, but it could mean brain damage. Of course, if the adoptive parents are willing to take the chance, they may take him home at their own risk in an independent adoption." He looked at me steadily. "The advantage of being adoptive parents is that they may choose," he said.

"But a slight temperature," I said. "How much of a risk is that?"

"Probably none," said the doctor. "If this weren't an adoption case we might not even mention it to the mother. Perfectly healthy babies sometimes have a little cold and run a temperature right after birth.

"The other alternative," said the doctor, fingering a paper on his desk, "would solve all your problems at once. You could sign here and release him to an adoption agency. They will be legally responsible for him and you don't have to give it another thought."

"And what would happen to him?" I felt suddenly very angry at this smug doctor who talked about not giving my child another thought.

"He would go to a foster home for six months," said the doctor. "Then if he is found to be healthy he will be placed for adoption. If he is not healthy, he will be placed in a proper institution under proper care and you will not have anything to concern yourself with in either case."

My hands were folded tightly in my lap and I heard the words come out of my mouth before I knew what I was going to say.

"It will be either the adoptive parents or myself," I said. "There can be no other alternative." As I heard my own voice, I knew it had to be that way. I couldn't send that little boy on to an uncertain future.

"You are talking about a full-time responsibility," said the doctor. "Are you prepared for all that that implies?"

"No," I said, "I'm not prepared, but my son needs a mother and he's going to have one—even if it's only me." The doctor looked at me for a long time. Then he put the form he was holding in his briefcase.

"I'll call the Carvers' pediatrician and inform him of your decision then," he said. "He will talk with the adoptive parents and you will know their decision tomorrow."

"Thank you," I said. "I will be waiting for their call."

On my way from the office I stopped in to look at my little boy. He was asleep in his crib, so unaware of the turmoil he is causing in several lives. But the turmoil is all of my making. I am responsible for bringing him into this world. Somehow I must see to it that he gets an even chance. If worst comes to worst and Martha and Don say no, I guess I can manage somehow. I think perhaps I am better prepared to be a mother now than I was months ago before I was willing to let him go out for adoption. When you give up your life it shall be given to you—isn't that what it says in the Bible somewhere?

But if he is to be mine, there is much to consider, a name to protect him, and my family. But now—not to think, not to get too involved, because surely Martha and Don will call to say they want him. Am I glad this is happening? I don't know. But at least I am kept from taking the easy way out, from just waking up after the delivery thinking that in seven days I will be putting on a new dress, new lipstick, and trying to forget.

Monday night

The nurse came to ask me if I would like to feed him. They all know what is happening and I can see in their faces their concern for the baby and for me. It was strange and wonderful to hold him at last, his body such a light weight on my lap, his head small enough to rest in the palm of my hand, little hands groping, little feet kicking. He has been so long inside me—and now to hold him, feel his soft skin smelling of baby powder. He drank all his milk and the nurse said that is good. Then she showed me how to burp him, how to support his back and neck, his downy head resting against my cheek.

He is so tiny, and yet a whole world in my arms. His eyes are drooping—he's so sleepy. Goodnight, baby. Tomorrow you go home with your mother.

Tuesday the 30th, Seventh Day

I am waiting for Martha's call. She should have called an hour ago. Dorothy will be here soon and we are scheduled to leave with the baby before noon. It is ten o'clock already. Is something wrong? I pray for the baby's sake that all is well. I don't worry about his health. I know he is all right. He must be. When I look at him, hold him, he is so alert, so quick to turn his eyes toward the light. When he is on his stomach, he tries to raise his head up high.

Is it too much for Martha and Don to take the chance? I know they must consider Kathy, too. Is it safer if he stays with me? I'm already attached to him. What if he goes with them and they find that he is sick? I don't know. I can only wait.

11:30 a.m.

The phone rang, but it wasn't for me. Hope rushed to my heart, and then came disappointment. I wonder what is keeping them? Have they talked to their doctor? Do they have to discuss it some more? Are they perhaps afraid of hurting me? How silly. They don't know how much they have done to make me grateful already. Whatever they decide won't change that.

I'm still wearing my housecoat. My things aren't packed yet. I feel exhausted. I wish they would call.

2:00 p.m.

I've talked to Dorothy on the phone and she has tried to call Martha and Don to find out what is happening. She called back to say their line is busy but she will keep on trying. She says not to worry, but how can I help it?

My little boy is fast asleep. I almost wish he were back inside me where nothing can happen to him. I feel so terribly helpless right now, and I shouldn't be, now that I have the responsibility for him also.

4:00 p.m.

Still no word from Martha and Don. We are supposed to leave here today and we'll have to go before evening. But I can't go before I know what is happening with Martha and Don. Dorothy called; she will be here before five o'clock. She will rent a motel room with a crib and a hot plate in case we need it. If we take the baby along, they'll give me twelve hours' supply of formula in four bottles and a complete layette with diapers and clothes. The nurse told me the Ladies' Auxiliary always gives a layette to the girls who keep their babies.

4:30 p.m.

Don called and his voice was so strained I almost didn't recognize it. "The doctor and the lawyer warn us against taking the baby home now. . . . I can't bear to see Martha suffer. . . . You understand, don't you?"

"Sure," I said. "I understand." I felt absolutely numb.

"Would you agree to have him placed in a foster home, under close observation by our pediatrician—we'll pay the bill—till we know for sure if he's all right?" Don sounded apologetic. "Martha doesn't want to do it that way but the doctor says we should."

"No!" I said, and the anger inside gave me strength to speak. "When that baby leaves here he goes with his mother—whether she is me or Martha. I chose you deliberately so that he wouldn't have to go to a foster home or an institution."

"I see," said Don in an empty sort of voice. "I guess there isn't anything more to say then, except thank you—and forgive us. We know this must be hard on you, Jean."

"Not as hard as on you," I said, and I meant it. "You lost a baby—I just got one." I put down the phone and went to lie on my bed. I could feel tears welling up but there was no time for them. I stared at the trees outside and tried to keep my mind a blank for just a minute, before the new thoughts came rushing in . . .

5:00 p.m.

The phone rang; it was Martha. "I can't bear it," she said. "If you don't want him in a foster home, I don't either. He needs a mother—and I need a baby. I don't care what happens." Her words came tumbling out. "I wouldn't have cared if he had been born to me. Please, Jean—will you let us have him?"

I could only feel a soreness inside—a horrible, dull ache. I saw the baby in his crib, so safe, so secure in his lack of knowledge of the world around him.

"What if he isn't all right?" I said.

"What if a son born to us hadn't been all right?" said Martha. "We will love him and give him what he needs, and we will never, never let him down."

"He needs you," I said, "more than he needs me. Please take him home." It was hard to keep from crying.

Martha said, "Now I feel that God is giving me a new chance to be a mother, for better or for worse." She was silent for a moment. "Don wants it this way, too, believe me, Jean. He just didn't want me to get hurt." Martha was crying openly now and laughing at the same time. "Oh, Jean," she said, "now I know how a mother feels. It isn't important that the baby suits us—but that we must try to be good for him."

"Thank you, Martha," I said, feeling a wave of exhaustion flooding me. "Your baby is ready to go home."

5:30 p.m.

I looked at him for the last time. He was asleep. I didn't want to hold him, and I suddenly remembered Carrie's words: "Don't hold him if you're going to give him away." But I'm not giving him away. He was never mine. I've just been responsible for him for a little while.

EPILOGUE

That was six years ago. Now I am married and have children of my own. Each day they grow to be more and more themselves, each one of them unique—each one wonderful, I think. And I am reminded of the lesson I learned several years ago; your children are not your children—they dwell in the house of tomorrow. What a blessing it is to be a guardian of their growth.

I know now also that a lesson once learned isn't always a lesson remembered. Only by self-discipline and will power does it become a lesson applied. My father once told me happiness comes not from doing what you like but from learning to like what you have to do. I know that now. I *try* to live it.

On her son's first birthday, Martha wrote Dorothy a card. He could walk then, and say a few words. "He is absolutely perfect in every way," wrote his mother.